RENEWABLE ENERGY: RESEARCH, DEVELOPMENT AND POLICIES

WIND FARMS

PERFORMANCE, ECONOMIC FACTORS AND EFFECTS ON THE ENVIRONMENT

Renewable Energy: Research, Development and Policies

Additional books in this series can be found on Nova's website under the Series tab.

Additional e-books in this series can be found on Nova's website under the eBooks tab.

RENEWABLE ENERGY: RESEARCH, DEVELOPMENT AND POLICIES

WIND FARMS

PERFORMANCE, ECONOMIC FACTORS AND EFFECTS ON THE ENVIRONMENT

MARIAN DUNN
EDITOR

nova publishers
New York

NOTICE TO THE READER

The Publisher has taken reasonable care in the preparation of this book, but makes no expressed or implied warranty of any kind and assumes no responsibility for any errors or omissions. No liability is assumed for incidental or consequential damages in connection with or arising out of information contained in this book. The Publisher shall not be liable for any special, consequential, or exemplary damages resulting, in whole or in part, from the readers' use of, or reliance upon, this material. Any parts of this book based on government reports are so indicated and copyright is claimed for those parts to the extent applicable to compilations of such works.

Independent verification should be sought for any data, advice or recommendations contained in this book. In addition, no responsibility is assumed by the publisher for any injury and/or damage to persons or property arising from any methods, products, instructions, ideas or otherwise contained in this publication.

This publication is designed to provide accurate and authoritative information with regard to the subject matter covered herein. It is sold with the clear understanding that the Publisher is not engaged in rendering legal or any other professional services. If legal or any other expert assistance is required, the services of a competent person should be sought. FROM A DECLARATION OF PARTICIPANTS JOINTLY ADOPTED BY A COMMITTEE OF THE AMERICAN BAR ASSOCIATION AND A COMMITTEE OF PUBLISHERS.

Additional color graphics may be available in the e-book version of this book.

Library of Congress Cataloging-in-Publication Data

ISBN: 978-1-63484-841-1

Published by Nova Science Publishers, Inc. † New York

CONTENTS

PREFACE

This book provides current research on the performance, economic factors and effects on the environment of wind farms. The first chapter provides a technical review of wind farm improved performance and environmental development challenges. Chapter Two explores a variety of methods to be used for assessing noise from wind farms. In Chapter Three, the potential impact of wind farms on radio devices in civil aviation and a review of the impact assessment procedure and methods of our research group is presented. Chapter Four discusses the measurement, analysis and improvement in the power quality of offshore wind farms.

Chapter 1 – The effective protection of the power converters of a Doubly Fed Induction Generator (DFIG) Variable Speed Wind Turbine (VSWT), could go a long way to improve its performance during transient conditions. A crowbar protection switch is normally used to protect the variable speed drive power converters during grid fault. The design of the pitch angle controller at the referenced coupled Rotor Side Converter (RSC) of the variable speed drive is also important in order to enhance its response during transient. This research work investigates the performance of a wind farm composed of variable speed drive considering five scenarios. In the first scenario, simulations were run for dynamic behaviour of a DFIG VSWT. The second scenario considers transient analysis for a severe 3LG fault. The third scenario shows the use of the crowbar switch to further enhance the performance of the DFIG VSWT in the second scenario. In the fourth scenario, a Flexible AC Transmission System (FACTS) device called Static Synchronous Compensator (STATCOM) was used to further enhance the stability of the variable speed drive. Finally, in the fifth scenario, a Current Controlled Voltage Source Converter (CC-VSC) was proposed to replace the conventional Voltage Controlled Voltage Source Converter (VC-VSC) used in the other scenarios. The simulated results show that the DFIG VSWT could perform better in all the scenarios based on the proposed protection and control techniques employed. Furthermore, some of the challenges of developing these variable speed wind farms ranging from environmental concern to government policies were also highlighted. Some opportunities were presented to make the establishment of these wind farms promising in the near future.

Chapter 2 – Wind farms have demonstrated impressive growth in electricity generation capacity over the past decades. Alongside this growth trend, some communities living in areas adjacent or close to existing and future wind farm sites have expressed concerns regarding possible health and environmental implications resulting from wind farm operations. Among the environmental concerns of wind turbine operations is the noise impact from wind farms. A wind farm operation should meet certain requirements in terms of noise impact. These noise limits are normally imposed by regulatory or planning authorities and are typically one of the strictest limits to be applied to potential noise sources. In many cases noise from wind farm operations is just above background or ambient noise present. Therefore monitoring and compliance checking of wind farm operation noise may be a complex scientific and engineering task. This chapter explores a variety of methods to be used for assessing noise from wind farms. The advantages and shortcomings of each approach to wind farm monitoring are discussed and considered within this chapter. Recommendations on implementations are provided based off the practicability and accuracy of results produced.

Chapter 3 – In recent years, with the quick development of offshore wind farms, there is an urgent and increasing demand on investigating the power quality of grid-connected offshore wind farm and understanding its impacts on the operation of power grid. This chapter focuses on addressing the aforementioned technical challenges and exploits the power quality issues of offshore wind farms from a number of aspects to enable us to model, analyze and protect the power quality of large-scale offshore wind farms. This chapter explores the modeling approach of semi-aggregated equivalent model of offshore wind farm based on PSCAD/EMTDC, which can be adopted for the study of measurement, analysis and improvement of power quality at point of common connection (PCC). Following to this, this chapter attempts to address this technical challenge through a simulation-based study by the use of PSCAD/EMTDC models and carries out an assessment of power quality at the Point of Common Coupling (PCC) in the scenario of offshore wind farm integrated into the power network whilst reduce the impact of index discrepancy and uncertainty. Finally, considering the integration of hybrid energy storage system (HESS) including battery energy storage system (BESS) and super-capacitors energy storage system (SCESS) to improve the power stabilization in power grid, the control strategy on managing the HESS to stabilize the power fluctuation in a real-time fashion without the need of predicting wind speed statistics is also presented. The suggested solutions are assessed through a set of simulation experiments and the result demonstrates the effectiveness in the simulated offshore wind farm scenarios.

Chapter 4 – Wind power is an attractive clean energy and wind farms increase with very high speed in recent years. However, as a particular obstacle, wind farms may degrade the performance of radio devices in civil aviation obviously. Therefore, wind farms may threaten the flight safety and correct impact assessment of wind farms on radio devices is important to guarantee the safety of civil aviation. In this chapter the potential impact of wind farms on radio devices in civil aviation and a review of the impact assessment procedure and methods of our research group is presented. The radio devices discussed in the chapter include surveillance devices such as primary surveillance radar (PSR) and second surveillance radar (SSR) and radio navigation devices such as very high frequency omnidirectional range (VOR) and instrument landing system (ILS). A wind farm usually comprises several wind turbines with very large size. The proper estimation of the scattering coefficient or radar cross section (RCS) of the wind turbine is of great importance to assess the impact of wind farms correctly. However, the intensity of electromagnetic scattering and the RCS of a wind turbine vary with several factors. Consequently a review of RCS estimation methods for a wind turbine of our research group is also presented in this chapter.

In: Wind Farms
Editor: Marian Dunn
ISBN: 978-1-63484-841-1

Chapter 1

TECHNICAL REVIEW OF WIND FARM IMPROVED PERFORMANCE AND ENVIRONMENTAL DEVELOPMENT CHALLENGES

K. E. Okedu*[1,2]*, R. Uhunmwangho*[2]*,
***Peter Ono Madifie*[2] *and C. C. Chiduole*[2]**
[1]Caledonian College of Engineering, Muscat,
Al Hail South, Sultanate of Oman
[2]University of Port Harcourt, Choba,
Rivers State, Nigeria

ABSTRACT

The effective protection of the power converters of a Doubly Fed Induction Generator (DFIG) Variable Speed Wind Turbine (VSWT), could go a long way to improve its performance during transient conditions. A crowbar protection switch is normally used to protect the variable speed drive power converters during grid fault. The design of the pitch angle controller at the referenced coupled Rotor Side Converter (RSC) of the variable speed drive is also important in order to enhance its response during transient. This research work investigates the performance of a wind farm composed of variable speed drive considering five scenarios. In the first scenario, simulations were run for dynamic behaviour of a DFIG VSWT. The second scenario considers transient analysis for a severe 3LG fault. The third scenario shows the use of the crowbar switch to further enhance the performance of the DFIG VSWT in the second scenario. In the fourth scenario, a Flexible AC Transmission System (FACTS) device called Static Synchronous Compensator (STATCOM) was used to further enhance the stability of the variable speed drive. Finally, in the fifth scenario, a Current Controlled Voltage Source Converter (CC-VSC) was proposed to replace the conventional Voltage Controlled Voltage Source Converter (VC-VSC) used in the other scenarios. The simulated results show that the DFIG VSWT could perform better in all the scenarios based on the proposed protection and control techniques employed. Furthermore, some of the challenges of developing these variable speed wind farms ranging from environmental concern to government policies were also highlighted. Some opportunities were presented to make the establishment of these wind farms promising in the near future.

1. INTRODUCTION

Energy conversion from wind into electrical energy system is rapidly growing because of the clean and renewable energy nature capability it possesses [1-3]. Speculations have it that by the end of 2020, the capacity of wind turbines that are going to be installed should hit 1900 GW [4]. Basically, a wind farm is a collection of various wind turbines of the same type or of different types to generate electricity.

Most of the wind turbine generators used in wind energy applications for sustainable energy production is fixed speed; however, the number of variable speed wind turbines (VSWTs) is on the increase by day [5-7]. The fact for the increase use of the VSWT is due to its ability to possibly track the changes in wind speed by shaft speed adapting; hence helps maintain optimal power generation. The control techniques of VSWT are very important and till date, more research is still going on in these areas. Principally, VSWT uses aerodynamic control systems like pitch blades or trailing devices that are variable in nature, but expensive and complex to achieve [8, 9].

The main aim of VSWT is power extractor maximization and in order to achieve this; the tip speed ratio of the turbine should be maintained constant at its optimum value despite changes of wind energy supplied. However, there exist mechanical and electrical constraints that are most common on the generator and the converter system. Therefore, regulation strategy of the effective power produced by VSWT is always one of the basic aims for the eminent and rapidly use of the turbine for energy production. Some of the merits of the VSWT over the Fixed Speed Wind Turbine (FSWT) are; cost effective, capability of pitch angle control, reduced mechanical stress, improve power control quality, improve system efficiency, reduced acoustic noise, etc. However, despite some of the above mentioned merits of the VSWT, there are some demerits of the wind turbine like fragile converter system that is vulnerable to damage during transient and also has a complex control topology.

In this study, the Doubly Fed Induction Generator (DFIG) is the VSWT. The DFIG possess reduction of inverter (20-30%) of the total energy system, potential to control torque and slight increase in efficiency of wind energy extraction [10]. However, the DFIG based VSWTs are very sensitive to grid disturbances especially to voltage dips. DFIG is made up of two converter control systems (rotor side converter and the grid side converter) which has a restricted over current limit, and needs special attention during transient conditions to avoid damage. When grid fault or transient occurs in the system, voltage dip is caused at the terminal voltage of the DFIG, consequently, the current flowing through the power converter may be very high current. In such situation, the conventional way could be to block the converters to avoid risk of damage because of their fragile nature, thereafter, disconnecting the generator and the wind farm from the grid. This act leads to the establishment of international grid codes. The grid codes require that wind turbine generators or wind farms must stay connected to the grid during grid fault or system disturbances and support or contribute to the network voltage and frequency. Thus, the DFIG based VSWT must comply with the Fault Ride Through (FRT) or Low Voltage Ride Through (LVRT) capabilities required by the grid codes. This practically means some requirements for the safe operation of the Rotor Side Converter (RSC) of the DFIG, because the rotor current and DC-link voltage of the wind generator will become very large during grid fault.

This work proposes a crowbar switch with effective resistance value to disconnect the RSC converter of the DFIG in order to protect it, thus operating the DFIG VSWT as a FSWT squirrel cage machine at transient conditions. As a further way of enhancing the DFIG capability, an investigation of different sizes of the crowbar switch resistor is necessary as different values of the crowbar resistor result in different behavior of the DFIG. Crowbar switch consist of set of thyristors or IGBTs that short circuits the rotor windings when triggered based on set optimal conditions. Consequently, the rotor voltage is limited, thus providing additional path for the rotor current, with improved DC-link voltage. Also, the output energy of the wind turbine depends on the methods of tracking the peak power points on the turbine characteristics due to fluctuating wind conditions [11].

An improved maximum power point tracking (MPPT) was employed in this work, whereby, the wind turbine is allowed to work with a speed close to its nominal value that permits the maximum power extraction. Thus, the pitch angle is kept constant at zero degree until the speed reaches a reference speed of the tracking characteristics. Beyond the reference point, the pitch angle is proportional to the speed deviation from the reference speed. In a bid to improve the performance of the VSWT, a detailed modeling of the turbine and its components were analyzed in this work. Different control strategies were employed ranging from the use of crowbar switch, FACTS device, different converter topologies (Voltage and Current controlled Voltage Source Converters) in addition to the MPPT tracking control system and pitch angle techniques. Simulations were run using the platform of Power System Computer Aided Design and Electromagnetic Transient including DC (PSCAD/EMTDC) visual environment. Dynamic (wind speed changes) and transient (grid fault) analyses were carried out to show the performance of the DFIG wind farm system respectively. Some challenges of siting the variable speed wind farm and some recommendations to enhance its effective operation were also given.

2. Review of Variable Speed Drives

The study of variable speed wind energy conversion system based on a doubly fed induction generator (DFIG) has been widely reported in the literature. Also, the Fault Ride Through (FRT) and Low Voltage Ride Through (LVRT) capabilities of this machine based on grid codes have been presented in the literature. References [12, 13], proposed sliding mode controls of active and reactive power of DFIG with MPPT for variable speed wind energy conversion. In these papers, the proposed control algorithm is applied to a DFIG whose stator is directly connected to the grid and the rotor is connected to the Pulse Width Modulation (PWM) converter. Wind energy integration for DFIG based wind turbines fault ride through and wind generation systems based on doubly fed induction machines were investigated in [2, 14, 15]. The authors did the FRT assessment of a DFIG and a Matlab Simulink for the DFIG variable speed wind turbine respectively. An MPPT using pitch angle with various control algorithms in wind energy conversion system was reported in [16] with the use of various intelligent control schemes in extracting maximum wind power using DFIG. Also, a sensorless MPPT fuzzy controller for DFIG wind turbine and hybrid sliding mode control of DFIG with MPPT using three multicellular converters were investigated and reported in [17, 18] respectively. It was concluded in the literature that the MPPT fuzzy logic control can capture the maximum wind energy without measuring the wind velocity and also that the DFIG MPPT connected by rotor side to three bridges of Multicellular Converters (MCCs), in conjunction with the Lyapunov stability method could improve the performance of the DFIG system during grid fault.

The integration of DFIG with a network having wind energy conversion system was carried out in [1], where two indirect converters associated with the principle of power distribution can operate the system conversion in a wide range of speed variation. DFIG with cycloconverter for variable speed wind energy conversion system for active and reactive power control was reported in [19]. In this paper, an MPPT control was included in the control system for improved performance of the DFIG system. Again, the modeling and MPPT control in DFIG based variable speed wind energy conversion system by using RTDS was investigated in [20], where the proposed control solution aims at driving the position of the operating point near the optimal set value.

The use of DC chopper, static series compensators, dynamic voltage restorer, Flexible AC Transmission Systems (FACTS) device, series dynamic braking resistor, super conducting fault current limiter, passive resistance network and antiparallel thyristors to improve the performance of DFIG VSWT have been presented in the literature. This work tends to review how the DFIG VSWT performance could be improved both in dynamic and transient conditions considering the magnitude of an active crowbar switch connected to the RSC of the machine, FACTS device and the power converter control topologies.

2.1. Crowbar Switch

Crowbar is made up of a symmetric three phase Y- connected resistance. It is usually connected to the rotor of the DFIG through a controllable breaker. In practice, the crowbar may be made up of one resistance fed through a switched rectifier bridge that would be sufficient to assess the overall impact of the crowbar protection on DFIG VSWT during transient. The breaker is normally open, but it is closed short-circuiting the rotor through the resistance if either the rotor current or the DC-link capacitor voltage becomes too high.

At the same time, the switching of the RSC Route Switch Controller is stopped [21, 22].

The value of the crowbar resistance is chosen according to [15, 23], as 20 times the rotor resistance. The choice of the crowbar resistance is important because it determines how much reactive power the DFIG will draw while the crowbar is inserted.

2.2. Concept of DFIG Pitch Control and MPPT

A lot of work have been done in the area of DFIG pitch control and MPPT techniques in the literature. However, how the crowbar switch affects some of the parameters of the DFIG VSWT would be investigated in this work. With the advancements in variable speed system design and control mechanisms of wind energy systems, energy capture and efficiency or reliability are paramount. Intelligent control techniques have been used to improve the performance and reliability of wind energy conversion system as reported in [24]. A further research was also carried out by same authors using fuzzy controller along with Hill Climbing Search (HCS) algorithm. As a brief, pitch control is the most common means for regulating the aerodynamic torque of the wind turbine and it works by searching for the peak power by varying the speed in the desired direction. The operation of the generator however, is in accordance with the magnitude and direction of change of active power.

The power gotten from wind energy systems depend on the power set point traced by MPPT. Tip Speed Ratio (TSR) affects the mechanical power from the wind turbine and thus, is defined as the ratio of turbine rotor tip speed to the wind speed. For a given wind speed, optimal TSR occurs during the maximum wind turbine efficiency. And in order to maintain this, the turbine rotor speed changes as the wind speed changes, thus extracting maximum power from wind. TSR calculation requires the measured value of wind speed and turbine speed data, but in the other hand, wind speed measurement increases the system cost and also leads to practical complexities.

2.3. Description of DFIG VSWT Simulation Models

One of the salient reasons for the wide use of the doubly fed wind induction generators connected to grid system is its ability to supply power at constant voltage and frequency, while the rotor speed varies. DFIG VSWT uses a wound rotor induction machine in addition to rotor winding supplied from the power/frequency converters. Thus, providing speed control together with terminal voltage and power factor control for the overall system. The use of the transient simulation analysis helps as a tool for the design of the rotor overcurrent protection and DC-link overshoot during transient. The overcurrent and over voltage protection mechanism in this study known as the crowbar, is used to protect the rotor side frequency converter during disturbances in the network [25, 26]. Space vector theory, based on model of a slip ring induction machine [27, 28] is the conventional approach to modeling the DFIG VSWT. This method provides sufficient accuracy also in cases when the voltage dips due to one or two phase faults in the network [29, 30]. The transient analyses of the DFIG wind turbine have been studied in [31, 32], where the crowbar switching is realized by using six anti-parallel thyristors.

2.4. Merits of VSWTs over FSWTs

Recent and sophisticated wind turbines are capable of flexible speed operations. The major advantages of flexible or adjustable speed generators compared to fixed speed generators are [33]:

- They are cost effective and provide simple pitch control; controlling speed of these wind generators allow the pitch control time constants to become much longer. Thus, reducing pitch control intricacies and peak power requirements. During cut-in or lower wind speed, the pitch angle is normally fixed. Pitch angle control is performed only to limit maximum output power at high wind speed.
- They reduce mechanical stresses; gusts of wind can be absorbed, i.e., energy is stored in the mechanical inertia of the turbine, creating an elasticity that reduces torque pulsations.
- The wind generators effectively compensate for torque and power pulsations caused by back pressure of the tower. This back pressure causes noticeable torque pulsation at a rate equal to the turbine rotor speed times the number of rotor wings.
- Improved power quality; torque pulsation can be reduced due to elasticity in the wind turbine system. This eliminates electrical power variations, i.e, less flicker.

- System efficiency enhancement; turbine speed is adjusted as a function of wind speed to maximize output power. Operation at maximum power point can be realized over a wide power range.
- Reduction in acoustic noise, due to lower speed operation is possible at lower conditions.

2.5. Advantages of DFIG in Wind Turbine Systems

DFIG phasor model is the same as the wound rotor asynchronous machine with the following two key points of difference [14]:

- Only positive sequence is taken into account, the negative has been eliminated.
- A trip input has been added. When this input is high, the induction generator is disconnected from the grid and from the rotor converter.

The basic advantages of DFIG in wind turbine system are as follows:

- The active and reactive power independent control through rotor current.
- Achieving magnetization of the generator through the rotor circuit and not basically through the grid.
- DFIG has ability to produce reactive power that is injected via the grid side converter (GSC).
- The converter size is normally 20-30% of the rated DFIG machine and is not based on the total power of the wind generator but on the speed range of the machine and therefore the slip range.
- Based on the economical optimization and increased performance of the system, the chosen speed range is decided accordingly.

3.1. Wind

Wind effect is one of most vital factors in modeling wind turbines. Wind models describe wind fluctuations in wind speed, which causes power fluctuation in generators. Basically, four components are paramount in describing a wind model [34] as shown below:

$$V_{wind} = V_{bw} + V_{gw} + V_{rm} + V_{nm} \tag{1}$$

where, $V_{bw}, V_{gw}, V_{rm}, V_{nm}$ are the Base wind, Gust wind, Ramp wind and Noise wind components respectively in (m/s). The base component is a constant speed; wind gust component could be described as a sine or cosine wave function or combination; a simple ramp function and a triangular wave may describe the ramp and the noise components respectively. The wind speed used in this study shows some of the wind components described above for the dynamic analysis of the system. A fixed wind speed was used for the transient analysis, because it is assumed that the wind speed did not change dramatically during the short time interval when the grid fault occured.

3.2. Equivalent Circuit of DFIG System

Figure 1 shows the equivalent circuit of the DFIG system. Due to its simplicity for deriving control laws, the Γ representation of the induction generator model will be used. It is important to note that from a dynamic point of view, the rotor and the stator leakage inductance have the same effect. Therefore, it is possible to use a different representation of the park model in which the leakage inductances are placed in the rotor circuit, the so called Γ representation of the induction machine [35]. The name is due to the formation of a shape like`Γ`of the inductances as shown in Figure 2. This model is described by the following space-vector equations in stator coordinates [36].

$$V_s^s = R_s i^s{}_s + \frac{d\Psi^s{}_s}{dt} \tag{2}$$

$$V_R{}^s = R_R i_R{}^s + \frac{d\psi^s{}_R}{dt} - j\omega_r \Psi^s{}_R \tag{3}$$

Subscript s indicates stator coordinates. The model can also be described in synchronous coordinates as

$$V_s = R_s i_s + \frac{d\psi_s}{dt} + j\omega_1 \psi_s \tag{4}$$

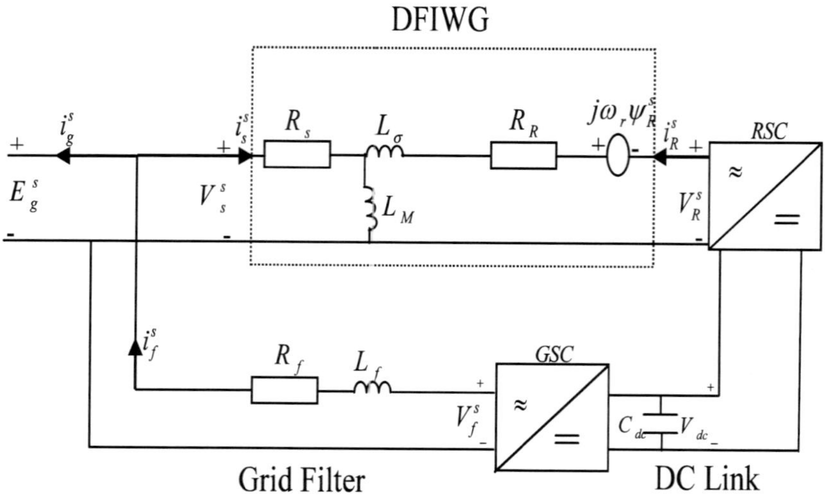

Figure 1. Equivalent circuit of DFIG system.

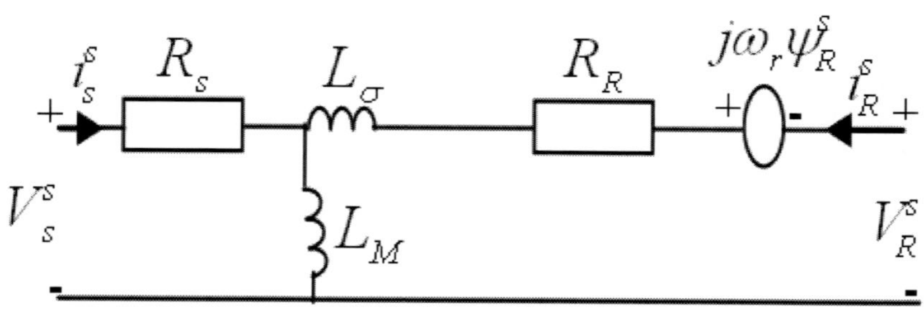

Figure 2. Γ representation of induction generator referred to the reference frame of the stator of DFIG.

$$V_R = R_R i_R + \frac{d\psi_R}{dt} + j\omega_2 \psi_R \tag{5}$$

where:

V_s Stator voltage;

V_R Rotor voltage;

i_s Stator current;

i_R Rotor current;

ω_1 Synchronous frequency;

ψ_s Stator flux

ψ_R Rotor flux

R_s Stator resistance

R_R Rotor resistance

ω_2 Slip frequency

The stator flux, rotor flux and electromechanical torque are given by

$$\psi_s = L_M(i_s + i_R) \tag{6}$$

$$\psi_R = (L_M + L_\sigma)i_R + L_M i_s = \psi_s + L_\sigma i_R \tag{7}$$

$$T_e = 3n_p I_m[\psi_s i^*_R] \tag{8}$$

where L_M is the magnetizing inductance, L_σ is the leakage inductance, and n_p is the number of pole pairs. Finally, the mechanical dynamics of the induction machine are described by [36, 37].

$$\frac{Jd\omega_r}{n_p dt} = T_e - T_s \tag{9}$$

where, J is the inertia and T_s is the shaft torque. The quantities and parameters of the Γ model relate to the park model (or the T representation) as follows:

$$V_R = \gamma V_R \tag{10}$$

$$i_R = \frac{i_r}{\gamma} \tag{11}$$

$$\psi_R = \gamma \psi_r \tag{12}$$

$$\gamma = \frac{L_{s\lambda} + L_m}{L_m} \tag{13}$$

$$R_R = \gamma^2 R_r \tag{14}$$

$$L_\sigma = \gamma L_{s\lambda} + \gamma^2 L_{r\lambda} \tag{15}$$

$$L_M = \gamma L_m \tag{16}$$

Grid Filter Model

Figure 3 shows the equivalent circuit of the grid filter. It is made of an inductance L_f and its resistance R_f. Application of Kirchhoff's voltage law to the equivalent circuit gives the synchronous coordinates model [35] in eqn. 17.

$$E_g = -(R_f + j\omega_1 L_f) i_f - L_f \frac{di_f}{dt} + V_f \tag{17}$$

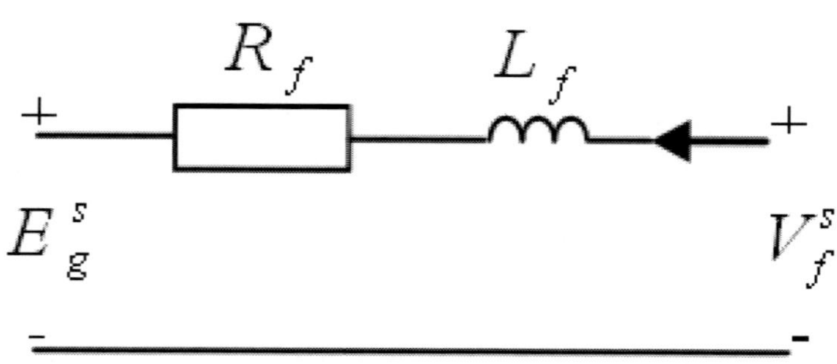

Figure 3. Grid-filter model of DFIG.

In eqn. 17 above, E_g is the grid voltage, i_f is the grid-filter current, and V_f is grid-filter voltage supplied from the grid-side converter.

3.3. Modeling of DFIG Wind Turbine

For electrical analysis, it is more convenient to use a simplified aerodynamic model of wind turbine as described by the set of equations in this section. The modeling of wind turbine is based on the steady-state power characteristics of the turbine. Thus, in order to effectively simulate the dynamic behavior of the wind turbine, the torque that it exerts on the mechanical shaft must comply with the following equation below [14]:

$$T_{turbine} = T_{em} = \frac{P_m}{\Omega_t} \tag{18}$$

Where P_m is the output power of the turbine, which is the mechanical power extracted from the wind, and is given as [36, 37]:

$$P_m = 0.5\rho A C_p(\lambda, \beta) v_{wind}^3 \tag{19}$$

Where, ρ is the air density in kg/m^3
A is turbine swept area m^2
C_p is the performance coefficient of the wind turbine
V_w is wind speed in m/s
λ is the tip speed ratio of the rotor blade tip speed to wind speed
β is the blade pitch angle in deg.
Ω_t is the mechanical speed of the wind turbine in rad/s

$$\Omega_t = \frac{\lambda v_{wind}}{R_t} \tag{20}$$

Also, the torque coefficient is expressed as

$$C_t = \frac{C_p}{\lambda} \tag{21}$$

Thus, the mechanical shaft could be defined as

$$T_{em} = 0.5\, \rho \pi R_t^3 C_t V_{wind}^2 \tag{22}$$

C_p in equation 19 is the power coefficient which can be expressed as a function of the tip speed ratio and pitch angle given by:

$$C_p(\lambda,\theta) = 0.22(\frac{116}{\lambda_i} - 0.4\theta - 5)e^{\frac{-12.5}{\lambda_i}} \quad (23)$$

$$\lambda_i = \frac{1}{(\frac{1}{\lambda + 0.08\theta} - \frac{0.035}{\theta^3 + 1})} \quad (24)$$

3.4. Control Strategies of Variable Speed Wind Turbine

The operating principle of the power flow for the DFIG system is explained as follows:

The mechanical power and the stator electric power output are defined by [36-38],

$$P_m = T_m \omega_r \quad (25)$$

$$P_s = T_{em} \omega_s \quad (26)$$

For a loss less generator, the mechanical equation is,

$$J\frac{d\omega_r}{dt} = T_m - T_{em} \quad (27)$$

For a loss less generator and in steady state at fixed speed, we have

$$T_m = T_{em} \quad (28)$$

$$P_m = P_s + P_r \quad (29)$$

It means that

$$P_r = -sP_s \quad (30)$$

where,

$$s = \frac{\omega_s - \omega_r}{\omega_s} \tag{31}$$

is defined as the slip of the generator.

Generally, P_r is only a fraction of P_s that is the absolute value of slip is much lower than unity and the sign of P_r is opposite to the slip sign. P_r is transmitted to or taken out of DC bus capacitor. The control of grid converter permits to generate or absorb the power in the grid in order to keep the DC link voltage constant. However, in steady state, for a loss less converters, the grid power is equal to P_r.

The converters have the capability of generating or absorbing reactive power and could be used to control the reactive power or the voltage at the grid terminals. The Rotor Side Converter (RSC) is used to control the wind turbine output power and the voltage (or reactive power) measured at the grid terminals. The Grid Side Converter (GSC) is used to regulate the voltage of the DC link bus capacitor. It is also used to generate or absorb the reactive power.

3.4.1. Active and Reactive Power Control of VSWT

The power is controlled in order to follow a pre-defined power speed characteristics. An example of such a characteristic showing also tracking of the reference power is shown in the MPPT nature in Figure 4 below [39]. The actual speed of the turbine ω_r is measured and the corresponding mechanical power of the tracking characteristic is used as the reference power for the power control loop.

We can see from the red line in the figure, the tracking characteristic is the locus of the maximum power point of the turbine (MPPT) i.e., maximum of the turbine power versus turbine speed curves.

For the power control loop, the actual electrical output power, measured at the grid terminals of the wind turbine, is added to the total power losses (mechanical and electrical) and is compared with the reference power obtained from the tracking characteristics. A Proportional Integral (PI) regulator is used and its output is the reference rotor current that must be injected in the rotor by the rotor converter. This is the current component that produces the electromagnetic torque T_{em}.

The reactive power at grid terminals or the voltage is controlled by the reactive current flowing in the rotor converter. When the wind turbine is operated in var regulation mode, the reactive power at grid terminals is kept constant by a var regulator.

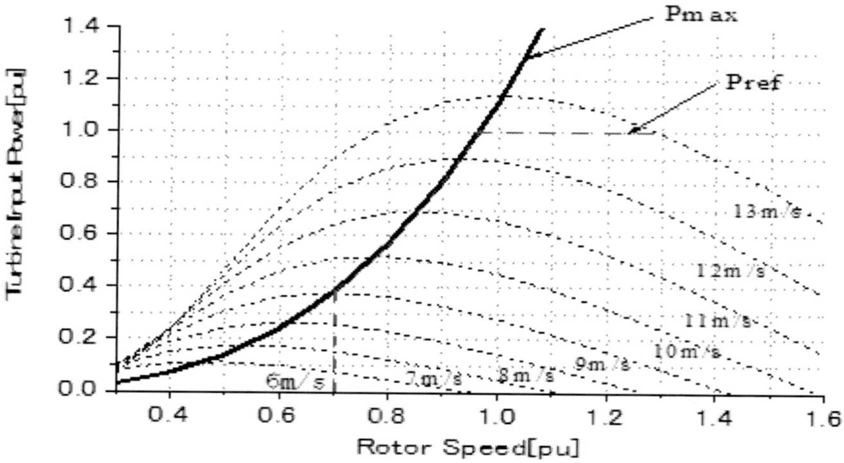

Figure 4. Turbine Characteristic with Maximum Power Point Tracking.

The output of the voltage regulator or the var regulator is the reference d-axis current that must be injected in the rotor by the rotor converter. The same current regulator as for the power control is used to regulate the actual direct rotor current of positive sequence current to its reference value. The RSC ensures a decoupled active and reactive stator power control according to the reference torque delivered by the MPPT control. The GSC controls the power flow exchange with the grid via the rotor, by maintaining the DC link bus voltage at constant by imposing the reactive power at zero.

The pitch angle is kept constant at zero degree until the speed reaches a point speed of the tracking characteristics. Beyond that point the pitch angle is proportional to the speed deviation from the point speed.

3.5. Detailed Structure of DFIG VSWT and Its Protection Schemes

The general block diagram showing the model of the DFIG system with its ancillary components and protection schemes, including power exchange of the converter systems and the grid system is shown in Figure 5.

The DFIG system in Figure 5 above is basically made up of two converter systems; the Rotor Side Converter (RSC) and the Grid Side Converter (GSC). A DC line with voltage V_{dc} is connected between both converter systems to store energy.

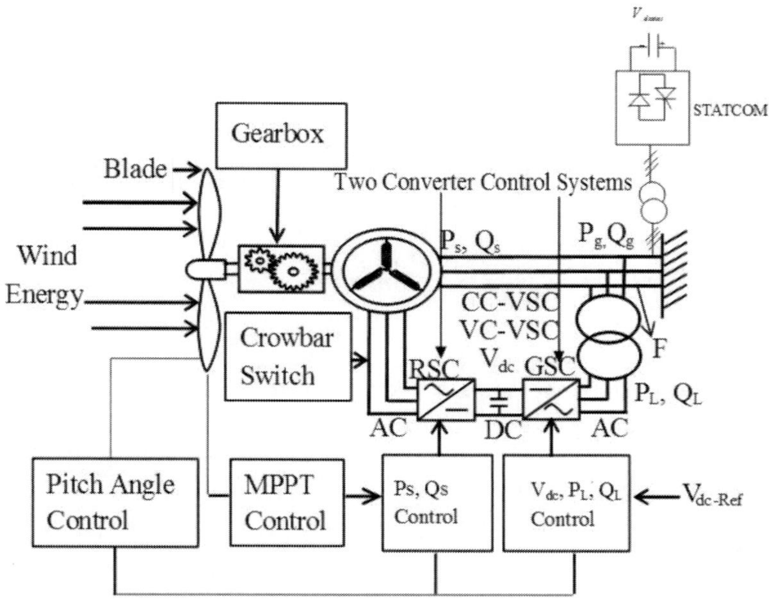

Figure 5. Protection Model for DFIG System.

The topology of the DFIG system is AC in the RSC to DC in the DC link voltage and again AC in the GSC system. The gearbox system in Figure 5 helps in matching the slow speed of the wind turbine to the very high speed of the generator system through proper mechanical coupling. The crowbar system is connected to the RSC and disconnects the RSC of the DFIG system during transient, thus making the DFIG behave like IG. After the fault is cleared as would be seen in the simulation results in the subsequent section, the crowbar reconnects the RSC of the DFIG system and normal operation of the wind turbine begins. The pitch angle control helps regulate the speed of the blade during high speed operation of the DFIG system against over speed to avoid damage of the wind turbine. It also reduces the risk of shutting down the wind turbine when high dynamics is experienced. The MPPT control helps regulate the obtained reference power of the wind turbine that is used to generate the reference switching voltages of the Pulse Width Modulation (PWM) in the RSC converter system. The RSC is controlled and coordinated by the MPPT reference power, active power of the stator P_s and the reactive power of the stator of the DFIG system Q_s. Also, the GSC of the DFIG system is controlled and coordinated by line power and reactive power at the grid side converter P_L and Q_L respectively, in addition to the DC-link voltage V_{dc} and its reference.

The description of the control systems in each unit of the DFIG shown in Figure 5 are discussed in the next session. A STATCOM external compensation device is also connected at the terminal of the DFIG system with the aim of providing further reactive power during transient. A Current Controlled Voltage Source Converter (CC-VSC) is further used to manipulate the DFIG control instead of the Conventional Voltage Controlled Voltage Source Converter (VC-VSC) system.

3.5.1. Rotor Side Converter (RSC) Control System

The block diagram for the RSC control system and its various control strategies is shown in Figure 6 below.

During steady state operation, the RSC initiates the electric power P_s and the absorbed reactive power Q_g by the DFIG. In Figure 6, the Phase Lock Loop (PLL) angle Θ_{PLL} is obtained from the generator rotor position as the effective angle for the abc-dq0 transformation for the reference currents and voltages obtained respectively. The difference between the stator power of the DFIG and the reference MPPT power obtained from the DFIG is passed through the first Proportional Integral (PI-1).

The tuning of the PIs used in this study is by the conventional trial and error method until the best results are obtained. The direct axis component current I_{dr} is used to regulate the generator power factor to unity, hence, the reactive power reference that would be absorbed Q_{ref} is equivalent to 0 value. This is done by using I_{dr} to control V_q^{r*}. Also, the absorbed reactive power at the grid connection end Q_g of the DFIG is compared to the reference and the error is sent to the third PI controller (PI-3) as shown in Figure 6, to evaluate the reference current I_{qr} for the quadrature component.

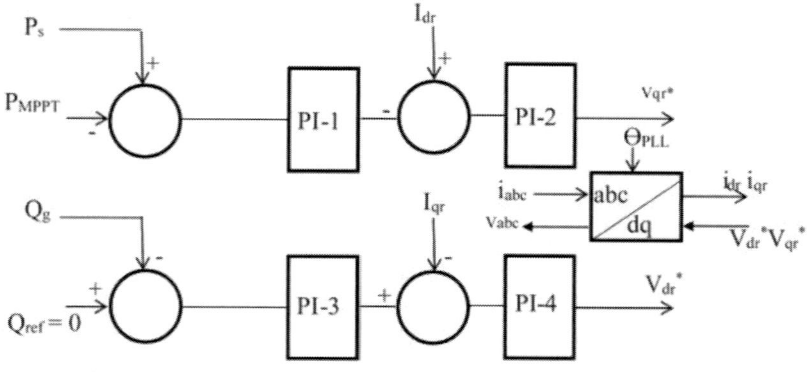

Figure 6. Rotor Side Converter of the DFIG Control System.

Therefore, I_{qr} is used to regulate V_{dr}^* after passing the signals through the two PIs (PI-3 and PI-4) respectively. A dq0-to-abc transformation, yields Vdr* and Vqr* respectively that are sent to the PWM signal generator.

Consequently, the three phase voltage desired V_{abc}^* are used for the switching of the Insulated Gate Bipolar Transistors (IGBT) in the converter system after comparing with a saw tooth generator system to generate pulses.

3.5.2. Grid Side Converter (GSC) Control System

The block diagram of the GSC control system and its control topology is shown in Figure 7 below. The Phase Lock Loop (PLL) angle used in the RSC is also used for the GSC as effective angle for abc-dq0 and dq0-abc transformations respectively.

The direct axis current component I_d is used to regulate the DC-link voltage V_{dc} to unity as per the reference DC-link voltage as shown in Figure 7. The difference in signal of the DC-link voltages is compared and is passed through PI-5. Thus, I_d regulates V_q^* while I_q regulates V_d^*. In normal operation the RSC already regulates the power factor of the DFIG, so there is no need for reactive power regulation by the grid side converter. Thus, $Q_{ref} = 0$ as shown in Figure 7, and is compared to the actual value of the grid side line reactive power Q_L and the error is sent to PI-7 controller to determine the reference current that is now passed through the summer to get the quadrature axis component. After a dq0- to-abc transformation, V_q^* and V_d^* are sent to the PWM signal generator. Thereafter, the IGBTs are switched using the generated V_{abc}^* reference signals in the PWM system.

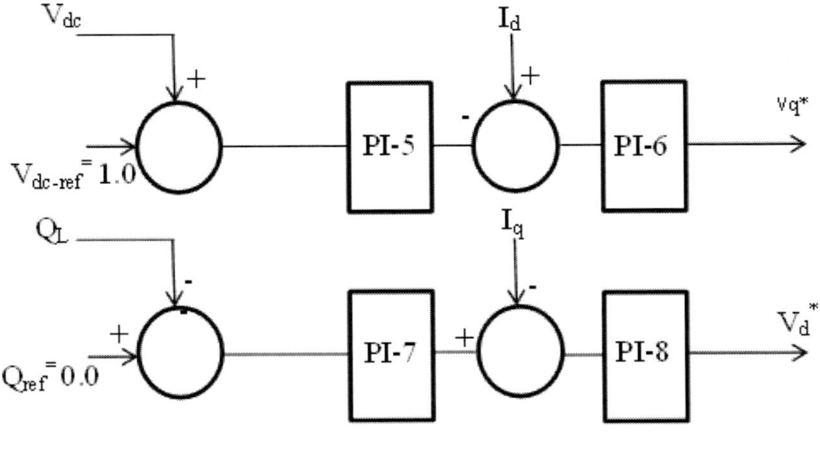

Figure 7. Grid Side Converter of the DFIG Control System.

3.5.3. Crowbar Control Strategy

The coordinated control of the crowbar control system and the RSC is shown in Figure 8 below. The crowbar system shown in the figure is used to assess the overall impact of protection of the DFIG during transient conditions. The crowbar is made up of one resistance fed through a three phase thyristors (T_1 to T_6) which is connected to the rotor side through a controllable three phase switch as shown in Figure 8.

The switch is normally open and is closed in order to short circuit the rotor, if the DC-link voltage exceeds the set maximum voltage of 150% pu during transient.

A comparator is used to compare the signals as shown in the figure. If V_{dcmax} is higher than V_{dc}, the comparator gives an ON state sign of 1 and thyristors (T_1 and T_6) are triggered, hence disconnecting the RSC and current flows through the resistance R_{cr} connected in the circuit. The value of the crowbar resistance has a great effect on the DFIG system during transient, as will be shown in the simulation results.

3.5.4. Pitch Angle System

The block diagram of the DFIG pitch angle control system used for this study is shown in Figure 9, where the pitch angle is kept constant at zero degree until the rotor speed of the DFIG VSWT reaches its limit speed of 1.3pu as shown in the wind turbine characteristics in Figure 4.

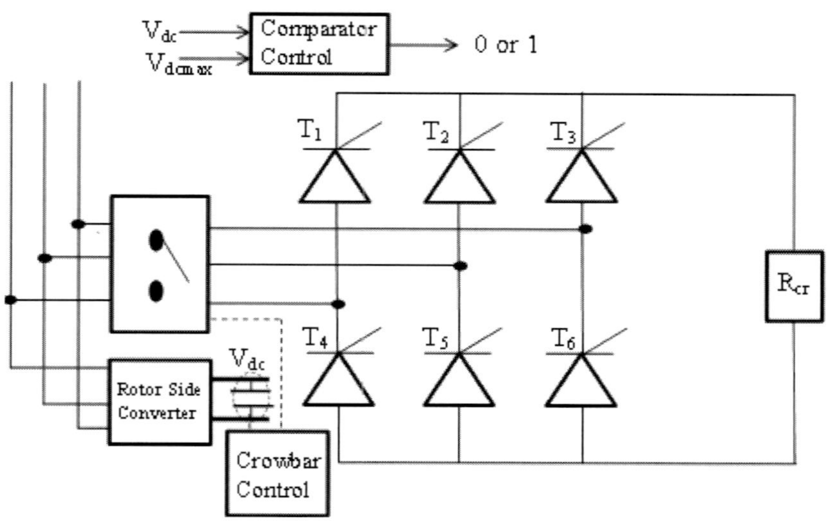

Figure 8. Crowbar Control Strategy.

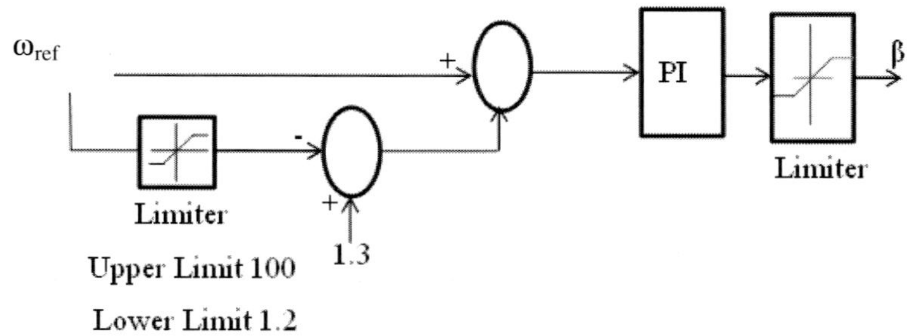

Figure 9. Pitch Angle Control System for DFIG Protection.

However, in order to improve the protection scheme of the DFIG wind turbine, which is the focus of this study, the pitching of the DFIG starts to operate when a speed of 1.2pu (prior to the limit of 1.3pu) is reached, via a limiter system. It would definitely help actuate fast the pitch angle controller to make for delay in the system operation during transient (grid fault) and dynamic (wind speed changes) conditions. The signal is passed through a Proportionate Integral (PI) system to generate an effective angle response β. For electromagnetic transients in power systems, the pitch angle control is of less interest. The rotational speed should be less than the speed value of 1.3pu. The simulation results showing the responses of the pitch angle controller during severe grid fault and high wind speeds will be discussed in the subsequent section of this work.

3.6. Parameters of the DFIG VSWT

The parameters of the DFIG VSWT used in this study are given as; rated power of DFIG 20MVA, rated voltage 690V, rated armature resistance 0.01pu, stator leakage resistance 0.15pu, magnetizing reactance 3.5pu, rotor resistance 0.01pu, rotor leakage reactance 0.15pu, inertia constant 1.5secs.

4.1. Analysis of Simulation Results

Some of the simulation results obtained using Power System Computer Aided Design and Electromagnetic Transient Including DC (PSCAD/EMTDC) [40] are shown in Figures 10 to 40. The simulation results are presented in five folds. The first fold is the response of the wind farm made up of DFIG VSWT during speed change. The second fold is the response during severe grid fault, while the third is the improvement of the DFIG response during grid fault by the active crowbar protection system. The fourth and fifth folds are the use of the STATCOM and the Current Controlled Voltage Source Converter control strategies respectively.

A. Wind Farm Response to Dynamics due to Wind Speed Changes

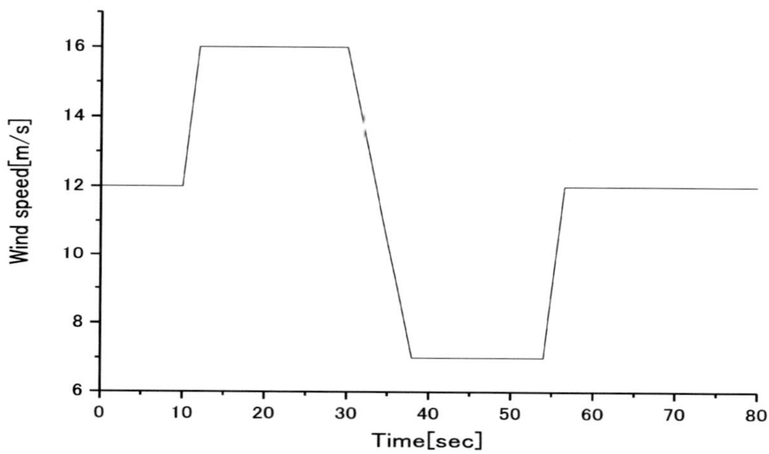

Figure 10. Wind Speed Pattern for the Study.

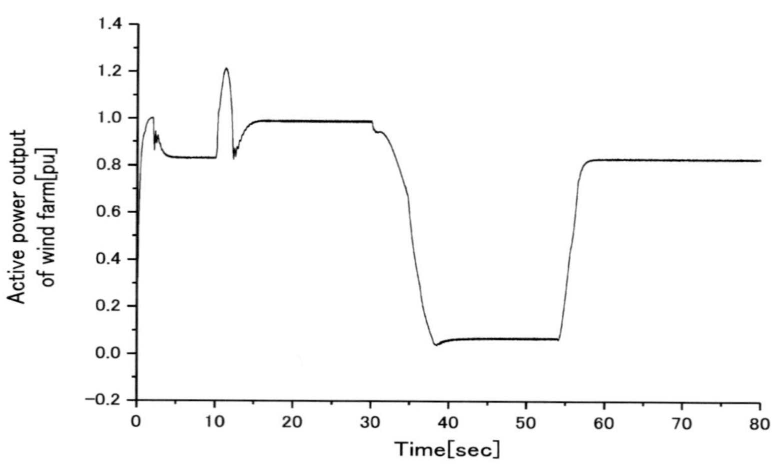

Figure 11. Active Power of the Wind Farm.

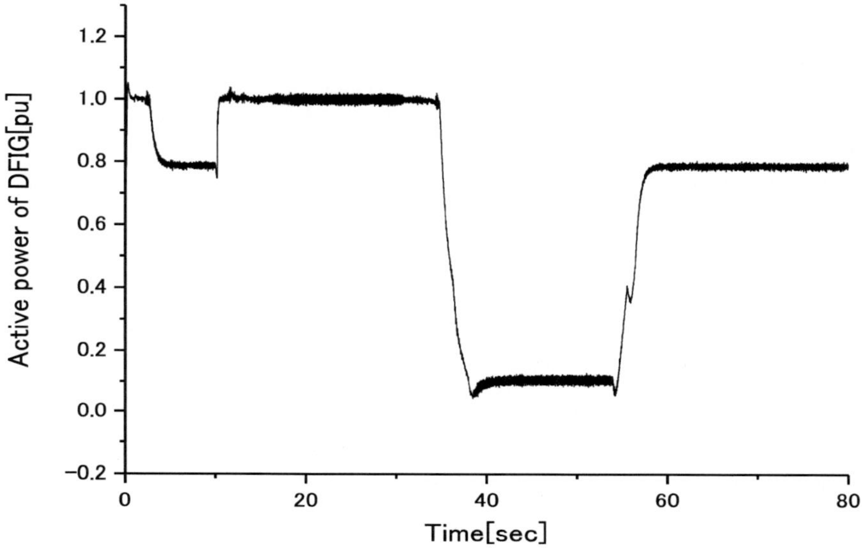

Figure 12. Active Power of DFIG.

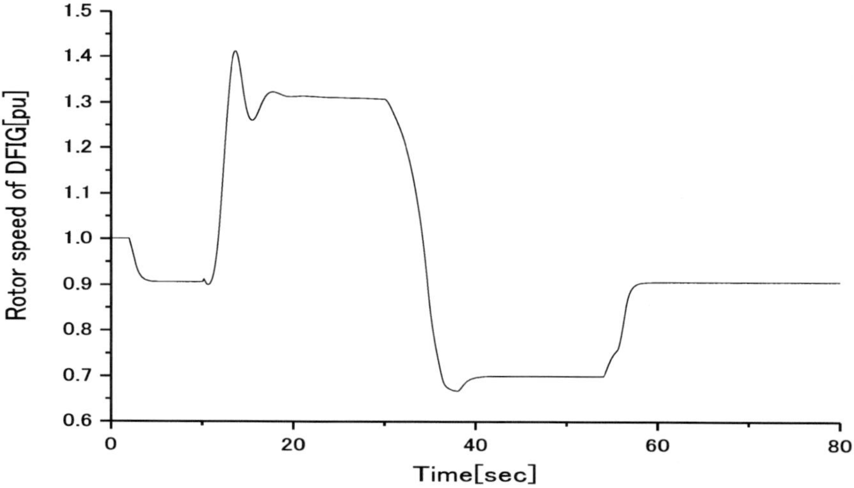

Figure 13. Rotor Speed of DFIG.

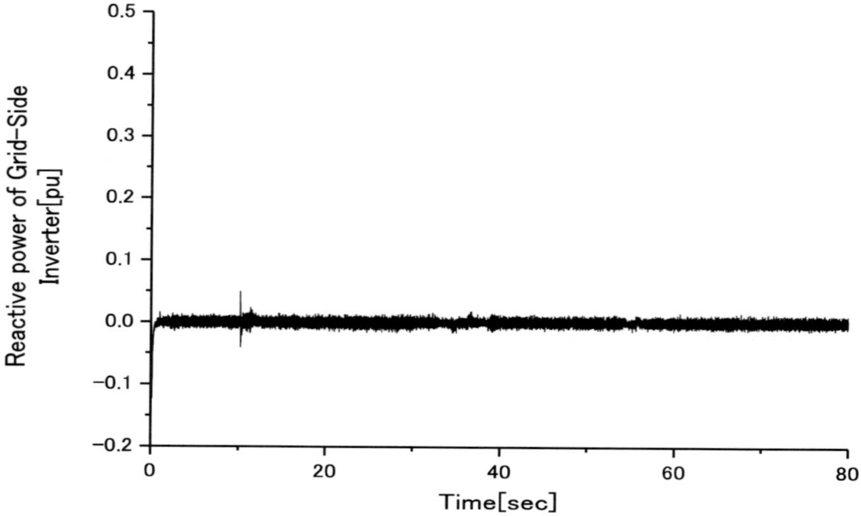

Figure 14. Reactive Power of DFIG GSC.

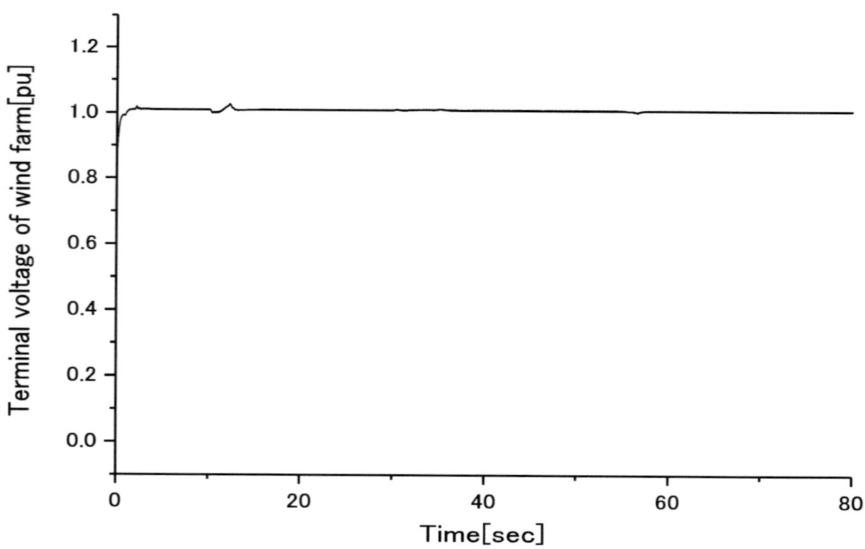

Figure 15. Terminal Voltage of Wind Farm.

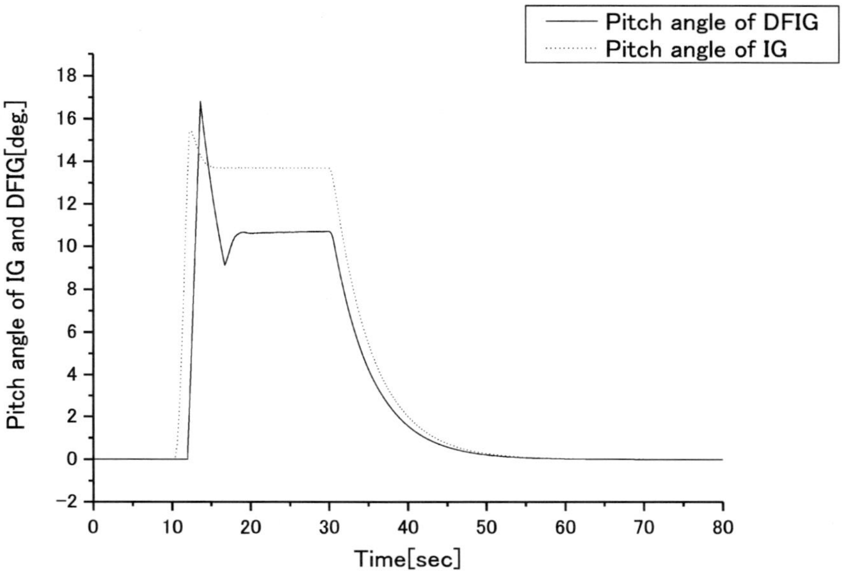

Figure 16. Pitch Angle of IG and DFIG.

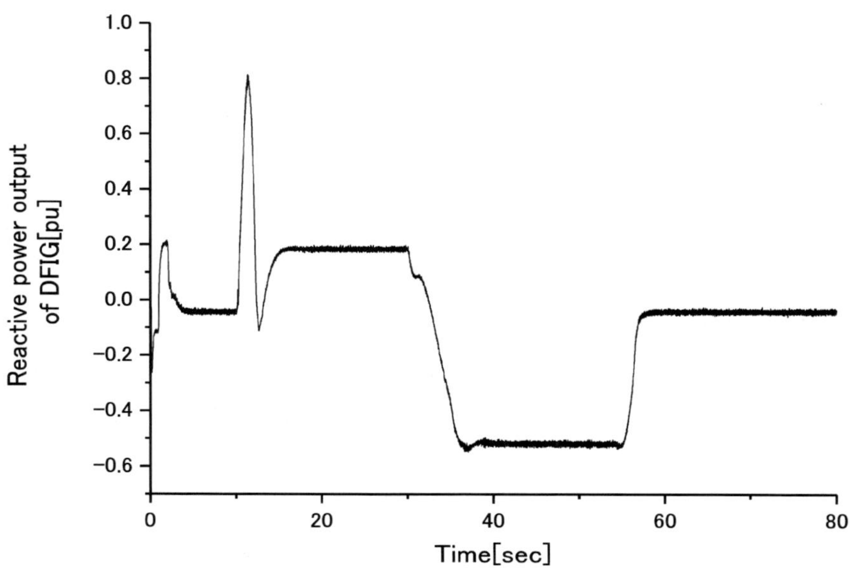

Figure 17. Reactive Power of DFIG.

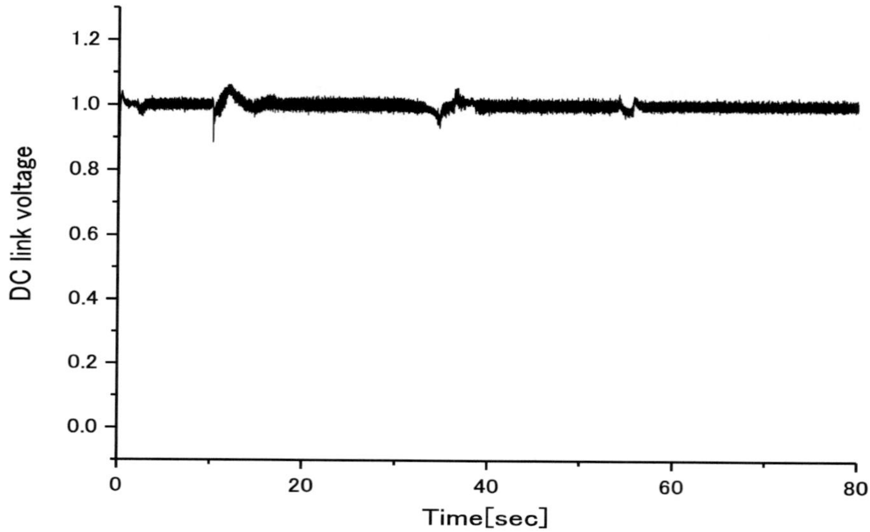

Figure 18. DC-link Voltage of DFIG.

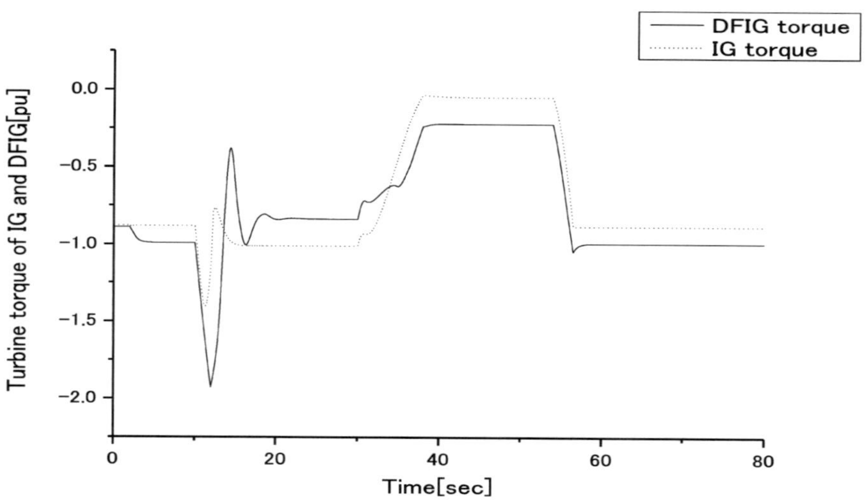

Figure 19. Turbine Torque of IG and DFIG.

B. Wind Farm Response to Transient due to Grid Fault

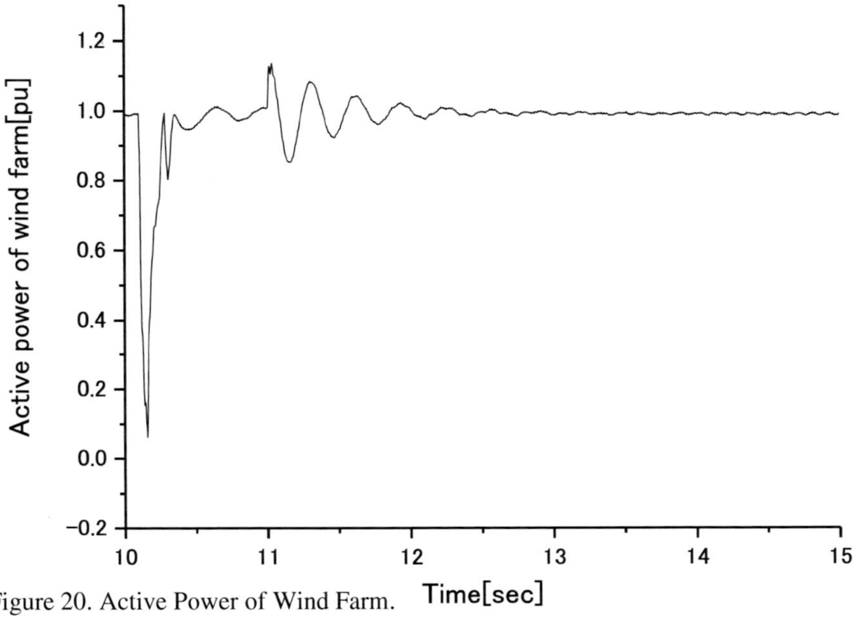

Figure 20. Active Power of Wind Farm.

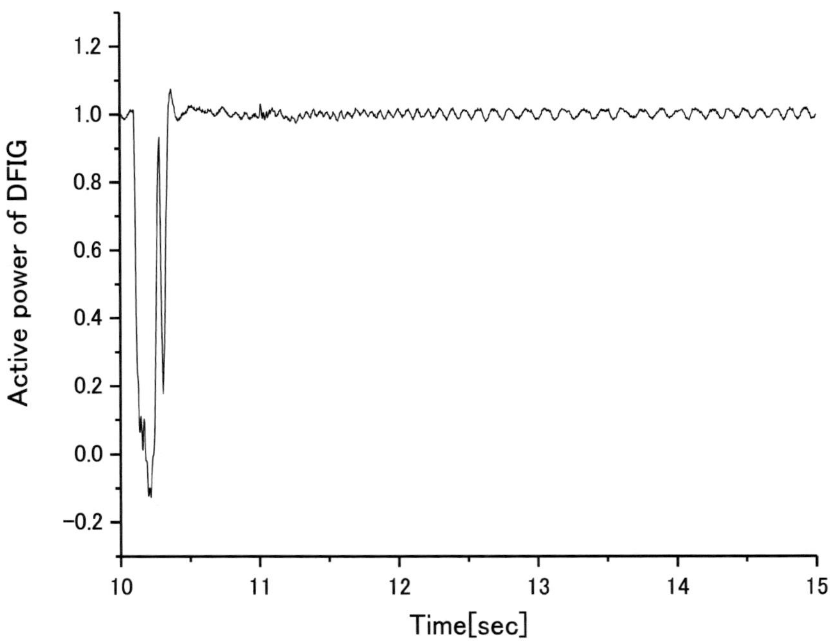

Figure 21. Active Power of DFIG.

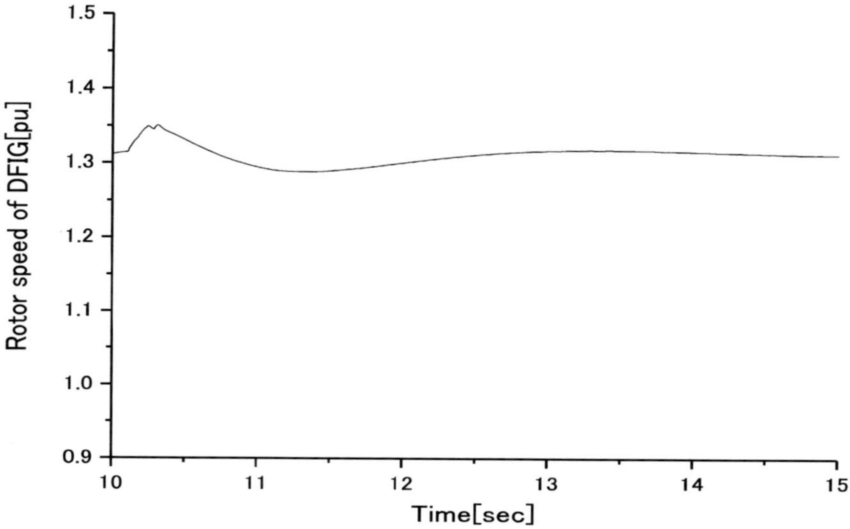

Figure 22. Rotor Speed of DFIG.

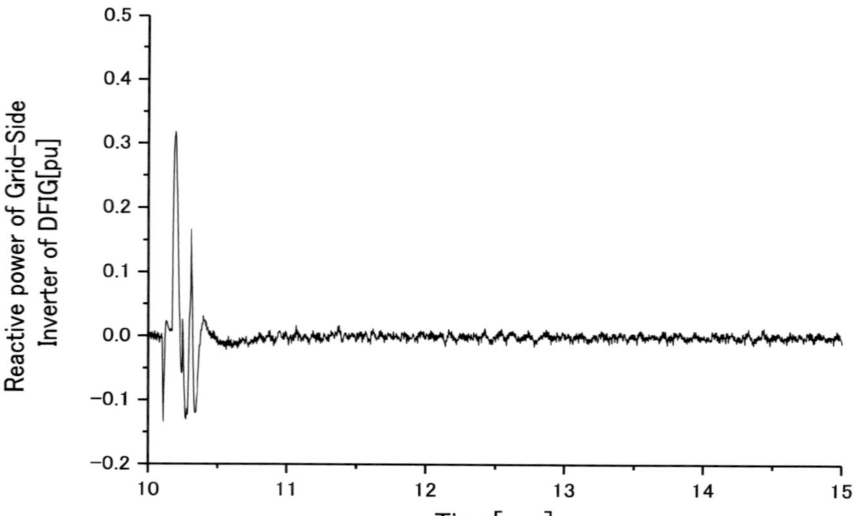

Figure 23. Reactive Power of GSC of DFIG.

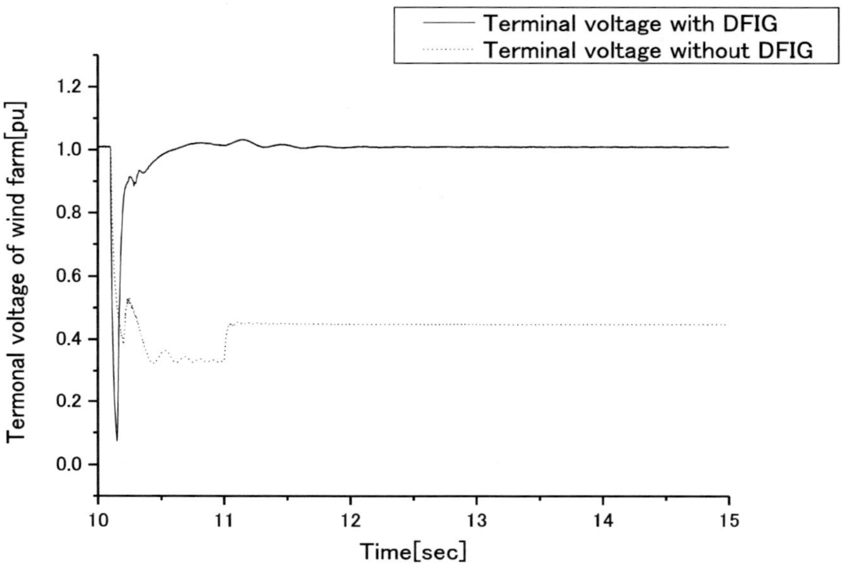

Figure 24. Terminal Voltage of Wind Farm.

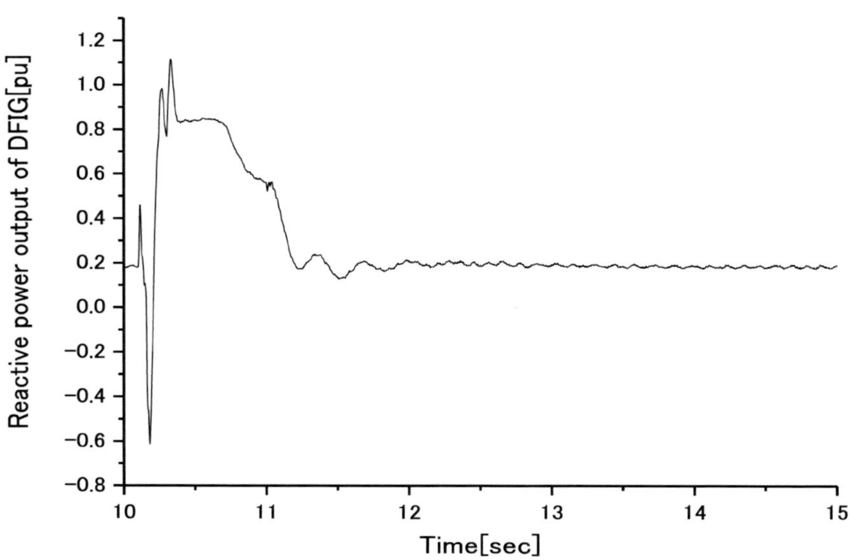

Figure 25. Reactive Power of DFIG.

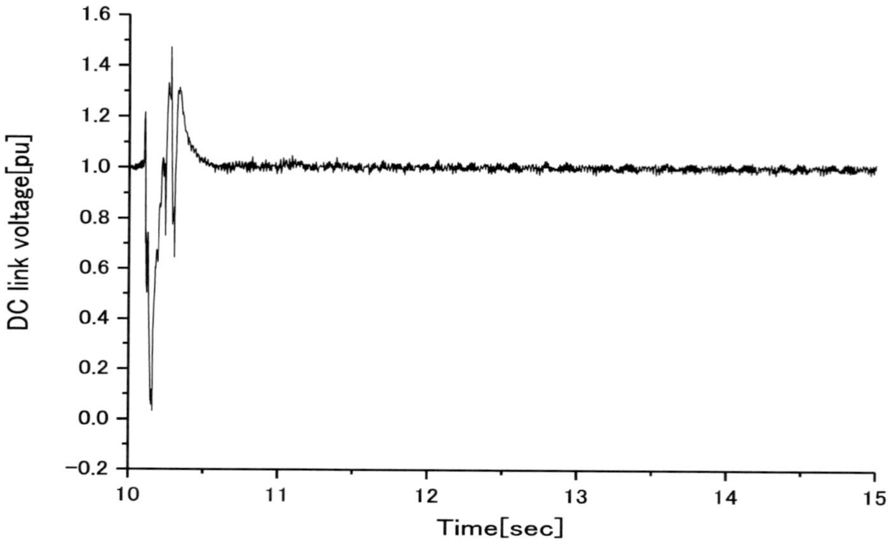

Figure 26. DC-link Voltage of DFIG.

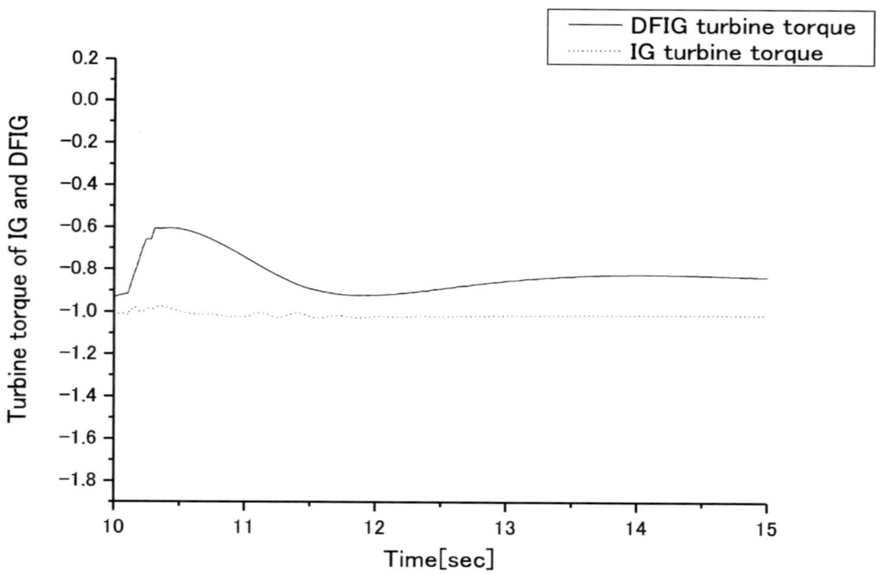

Figure 27. Turbine Torque of IG and DFIG.

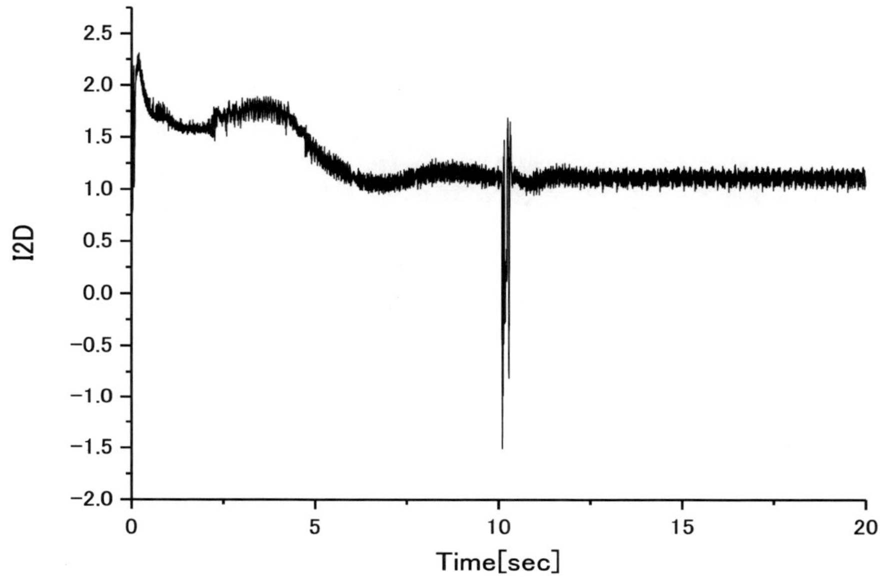

Figure 28. Reference d-axis Switching Current of DFIG.

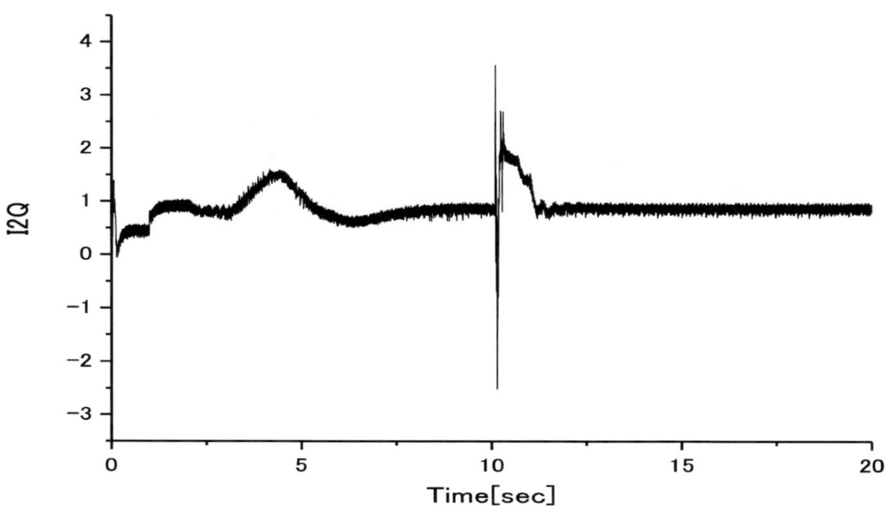

Figure 29. Reference q-axis Switching Current of DFIG.

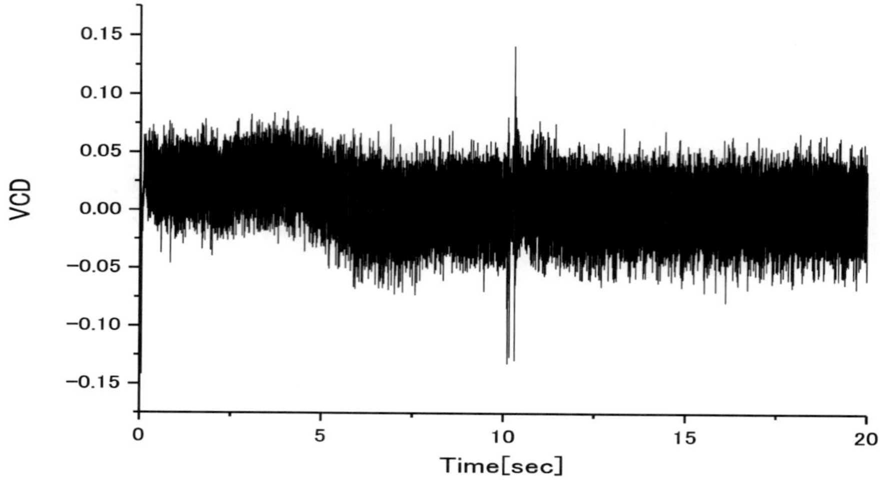

Figure 30. Reference d-axis PWM IGBT Switching Voltage.

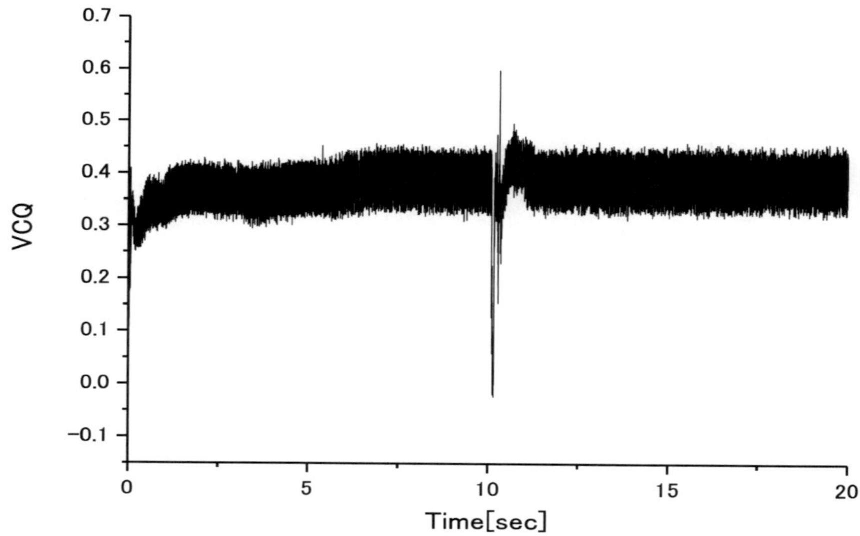

Figure 31. Reference q-axis PWM IGBT Switching Voltage.

C. Analysis with Crowbar Switch System

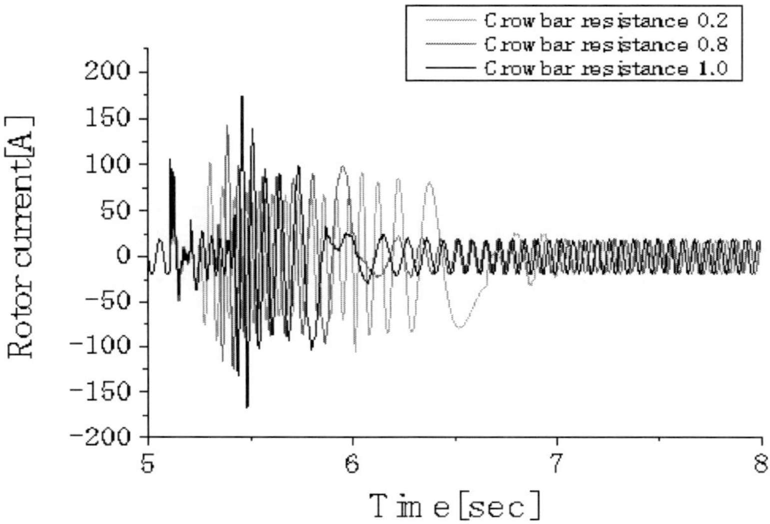

Figure 32. Rotor Current for Different Crowbar Resistances.

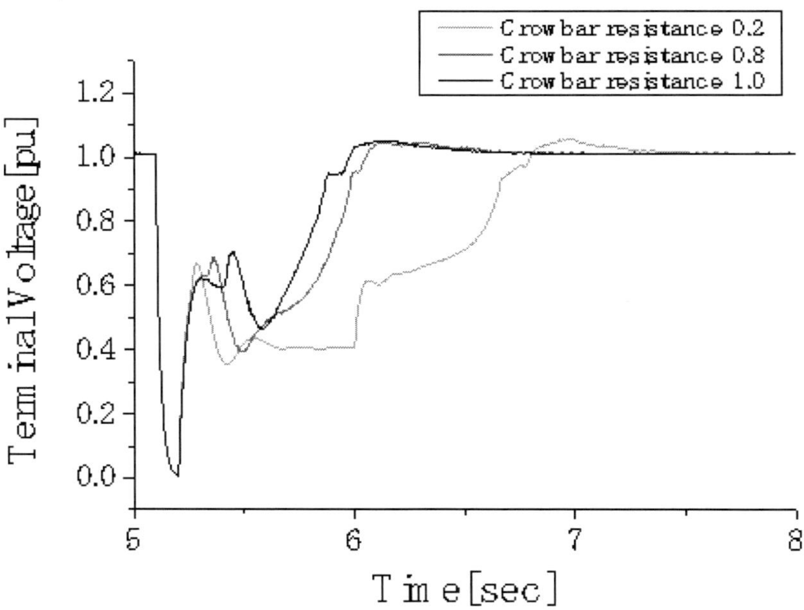

Figure 33. Terminal Voltage for Different Crowbar Resistances.

D. Analysis with FACTS (STATCOM) System

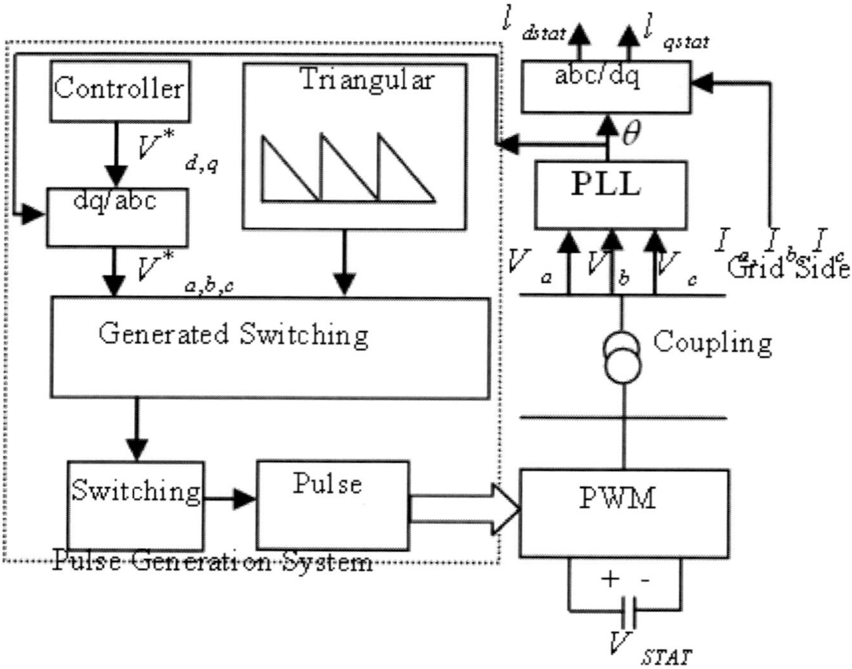

Figure 34. STATCOM Control Topology.

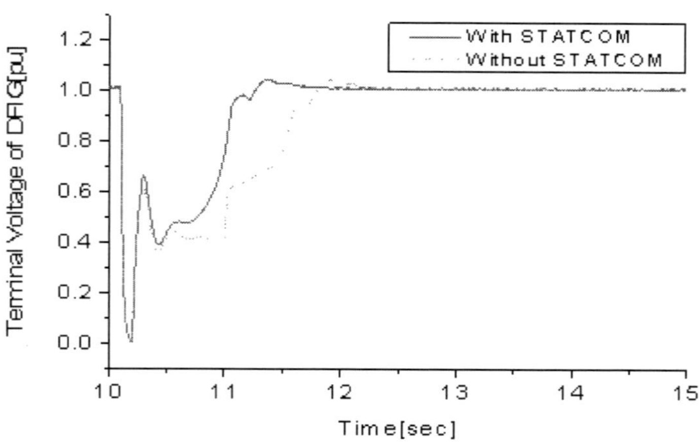

Figure 35. Response of Terminal Voltage of DFIG.

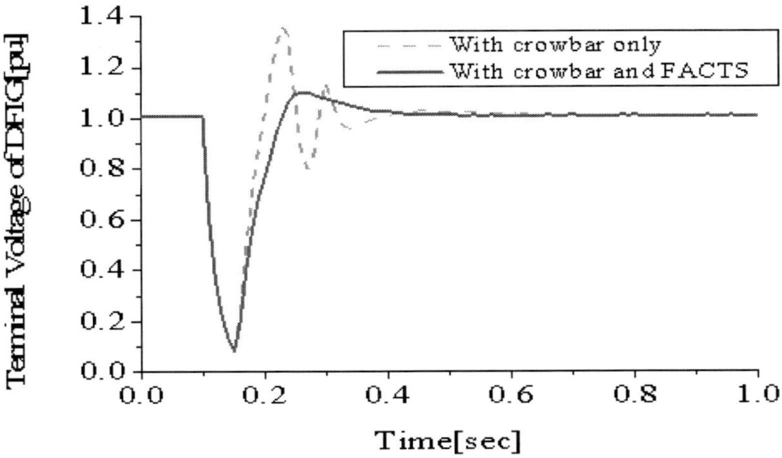

Figure 36. Response of Terminal Voltage of DFIG.

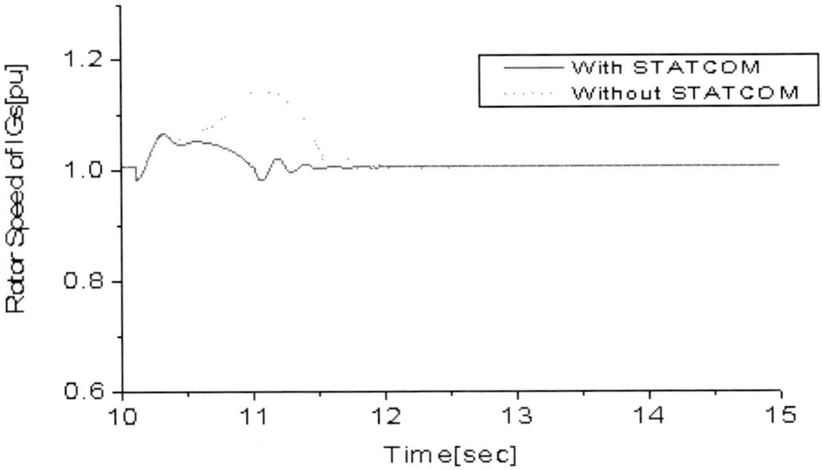

Figure 37. Response of Rotor Speed of IG.

E. Analysis with Current Controlled-Voltage Source Converter Topology

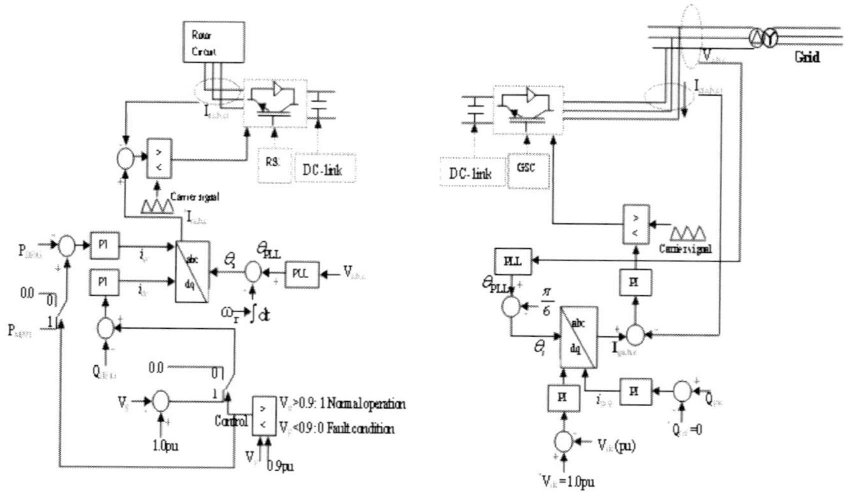

Figure 38. DFIG CC-VSC Switching Topology.

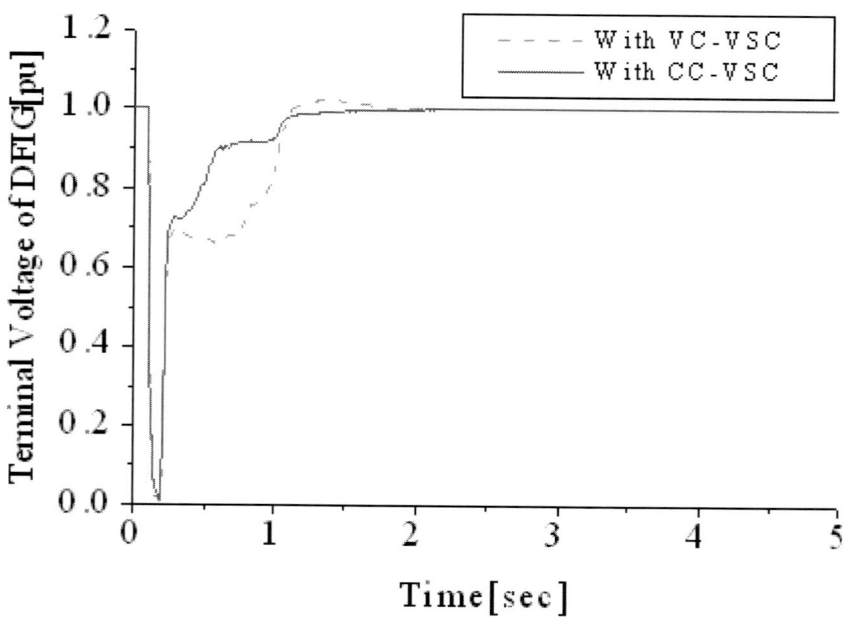

Figure 39. DFIG CC-VSC Terminal Voltage Response.

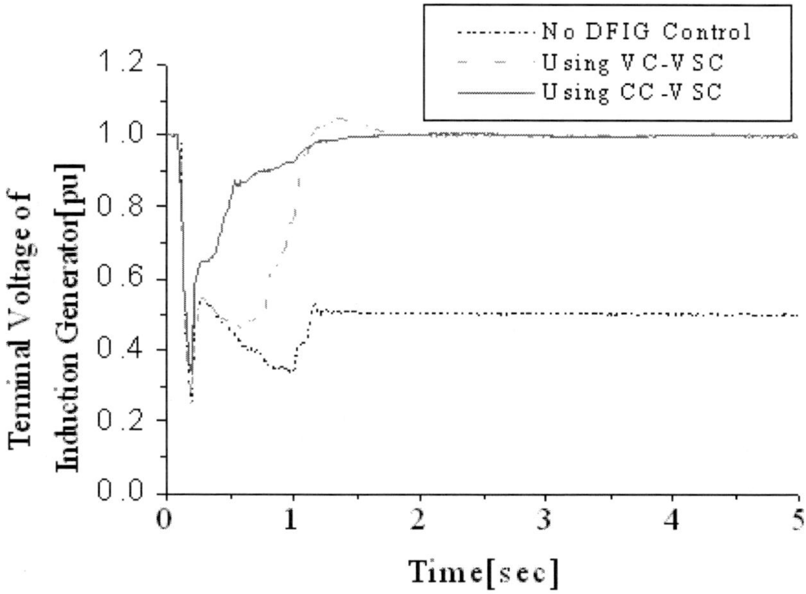

Figure 40. IG CC-VSC Terminal Voltage Response.

4.2. Discussion of Results

Figure 10 shows the wind speed pattern used in this study. The wind speed pattern is based on the wind components discussed earlier. The wind speed was constant at $12m/s^2$ and then ramped to $16m/s^2$. The cut-in and cut-out wind speeds from the figure are $7m/s^2$ and $16m/s^2$ respectively. The rated wind speed is $12.44m/s^2$. As part of one of the objectives of this work, to model the DFIG VWST wind farm and show its dynamic behavior during wind speed, it could be realized from Figure 11 to Figure 14, that the active power of the wind farm, the active power of the DFIG system and the rotor speed of the DFIG follow the wind speed pattern, showing the effectiveness of the controllers used for the DFIG study.

Due to the speed limit of 1.3pu set by the pitch controller of the DFIG in Figure 9 for periods of high winds during operation, the robustness and effective operation of the pitch controller could be seen in the response of the rotor speed of the DFIG system. The high speed effect led to overshoot in the rotor speed response, but is however brought to its 1.3pu limit of design after a short while as shown in Figure 13, thus protecting the DFIG and also helps avoid the wind farm shut down due to high wind speeds based on international grid code requirements.

In Figure 14, the reactive power of the DFIG VSWT system was maintained at 0 value based on the topology of the controller given in Figures 6 and 7 respectively. Since the reactive power is directly proportional to the voltage of the system, it could be seen that in Figure 15, the terminal voltage of the DFIG VSWT and hence the wind farm was maintained constant at 1.0 pu despite the wind speed change pattern. Again, these responses are confirming the effectiveness of the variable speed drive controller for reactive power provision. The pitch angle of the DFIG system as compared to that of the IG are shown in Figure 16, where it could be observed that the DFIG pitching is faster in response compared to the IG, due to the control system employed. The reactive power supply of the DFIG system for the wind farm stability control is shown in Figure 17. During very high and low wind speeds more reactive power is needed to achieve stability of the wind farm.

In Figure 18, the DC-link voltage is seen to be kept constant at 1.0pu based on the control strategy shown in Figure 7. Thus, the changes in wind speed at 10secs, 38secs and 56secs could respectively be observed in the specks of the DFIG DC-link voltage response. The turbine torque for the DFIG VSWT system and its response when it is used as IG system are given in Figure 19 for both systems respectively.

The wind farm responses for transient condition due to grid fault are given in Figures 20 to 31 for the second fold of results analysis in this study. In Figures 20-23, the active power of the wind farm, active power of the DFIG system and rotor speed of the DFIG, were able to recover after a severe 3LG fault that was applied at point F in the model system of Figure 5. This could be explained from the fact that the reactive power injected by the grid side converter in Figure 23 that is within the range of 20-30% of the DFIG capacity as discussed earlier in this work, was sufficient enough to improve the stability of the wind farm system. The effect of the reactive power injection by the GSC of the DFIG could also be seen in Figure 24, where the terminal voltage was able to recover after the severe fault and was not able to recover when the DFIG system control topology was not implemented. This voltage collapse scenario is typical for a squirrel cage Induction Generators (IG). Hence, the voltage collapse shown in Figure 24, would lead to shut down of the wind farm during transient conditions to avoid damage to utilities in particular and the environment in general.

Figures 25-27 show the responses of the reactive power, DC-link voltage, and turbine torques of the DFIG VSWT and IG responses to the severe grid fault. The responses of the various components of the DFIG VSWT switching system in the PWM converter are shown in Figures 28-31. These switching signals help in the control strategy of providing active and reactive power regulation for the wind farm to help achieve stability based on Figures 6 and 7 respectively. The third fold of the simulation analysis is concerned with the crowbar switch system of Figure 8. The responses of the DFIG VSWT due to the severe 3LG grid fault could be further enhanced using the crowbar switching topology. An investigation of the effects of the crowbar resistance sizing on the DFIG VSWT stability as earlier discussed was carried out. The simulation results as seen in Figures 32 and 33 respectively could help establish the fact that the crowbar switch has great influence on the stability of the DFIG system during grid faults. As seen in Figures 32 and 33 [41] for the DFIG rotor current and DC-link voltage respectively, when the value of the crowbar resistance is too small (0.2pu) or too large (1.0pu), the responses of the DFIG variables results in either overshoot or delay in settling time. However, when a proper crowbar switch sizing of 0.8p.u was used, the simulation results in Figures 32 and 33 show the effectiveness of the crowbar switch to help achieve stability of the DFIG system for sustainable supply of power to the utilities and thus help avoid the wind farm shut down which could be hazardous to the environment.

In the fourth fold of the result analysis, the STATCOM FACTS reactive compensation external device was used to enhance the variable speed drive [42-44]. The control strategy of the considered FACTS device is shown in Figure 34, while the responses for the DFIG terminal voltage with only the FACTS device considered is given in Figure 35 with improved performance during the grid fault. When the Crowbar system was considered along the FACTS device, a better performance was achieved during the grid fault as shown in Figure 36. Again, when the variable speed drive is tied to the IG system, an improved performance of its rotor speed is obtained as shown in Figure 37 because of the extra reactive power injected into the system. The last fold of results analysis in this work is when the conventional Voltage Controlled Voltage Source Converter is replaced with a Current Controlled Voltage Source Converter system [45, 46] as shown in Figure 38. Response of the DFIG terminal voltage with improved performance is shown in Figure 39 considering the CC-VSC topology. Figure 40 shows the response of the IG system when it is tied to the DFIG considering the VC-VSC, CC-VSC and the scenario when no control is implemented. In all these cases, the responses obtained using the CC-VSC is more favorable and cost effective because less circuitry is involved.

5. Challenges and Opportunities of Wind Farm Development

In light of the presented technical review of the behavior of the variable speed wind farms considering international grid codes to maintain continuity of supply during dynamic and transient conditions, it is pertinent to highlight some of the key challenges in their location and development with respect to environmental concerns. Some of the key challenges and opportunities in this regard are discussed as follows:

5.1. Noise and Safety

Wind turbines are mechanical devices; therefore, they are bound to be some noise during their operation due to the wind. This challenge could create opportunity for design changes in wind turbines to reduce noise in wind farms and the environment by making modern wind turbines convert more of the wind energy into rotational torque with less into acoustic noise. A further method of reducing wind turbine noise in wind farms could be by employing insulating materials and well planned siting of the wind farm.

Furthermore, since wind turbines are related to blade movement, it could be dangerous to the environment especially in accessing the public. Wind farms are normally sited far from residential areas in this regard.

5.2. Visual Effects

Since wind farms are sited in open places to capture wind energy, they are bound to be highly conspicuous and affect the aesthetic or beauty of the landscape. An opportunity in light of this could be to locate less number of wind turbines at a time in a particular site, thereby making multiple locations. Also, this effect could be mitigated by employing better or improved models of wind turbines that are more efficient and bigger.

5.3. Mortality of Wild Life

A biological and environmental issue pertaining to wind turbines in wind farms is the death of birds and bats, raising concern by conservation agencies. A good opportunity in this regard is to carryout proper studies taking into consideration wildlife animals like birds and bats with respect to evaluating measures and design methods to mitigate collision.

Another measure is a proper and careful site location of the wind farms to reduce wildlife mortality.

5.4. Interference

There is potential interference with radar and telecommunication facilities by wind turbines. There are also electric and magnetic fields in wind farms since they are electrical generating facilities. This has great influence in the design and viability for potential wind farms. Conflicts over airspace that are very densely populated are great challenges. This could be taken care of by proper siting of wind farms consideration the waves around the environment.

5.5. Logistics

As seen from the presented simulation results, the VWST DFIG based gives room for rapid progress in turbine technology design to optimize energy generation from low wind speeds for dynamic analysis. The effective prediction of energy production of large wind turbines on complex site remains a key challenge. Thus, the extension of the towers of wind turbines with very high towers comes with a high cost, and if it makes financial justification to build high and big wind projects. It is necessary to balance good wind resources against huge range of geo-economic factors like accessibility of site, transportation and ease of connection of wind farms to the grid, expertise and tools available for construction and effective maintenance, in order to evaluate the rate of return of investment for wind farm developers. Good wind sites are located in remote locations that are very far from the utilities. Hence, wind farm developers must provide expensive transmission lines to evacuate electricity to the end users. To overcome this problem, wind farm developers have to look beyond the cost of installation and logistics for large wind turbine.

5.6. Policy of Governments and Human Factors

Government policies and stability affects the effective and smooth running of wind farms. Most government see wind energy as unreliable, hence are not willing to invest in wind projects. Also, the protests and claims of the indigenes where wind farms are sited create a big challenge to wind farm development. Most people like the use of wind energy, but they do not want wind farms located close to them due to noise, visual effects and other challenges like health problems if they live close to wind farms. A way out of this problem could be to employ an approach tagged smart community engagement, whereby wind farm developers engage and enlighten the local people about the pros and cons of wind farms siting in their locality.

Conclusion and Recommendations

Conclusion

This work presented a mathematical model of a Doubly Fed Induction Generator (DFIG) Variable Speed Wind Turbine (VSWT) system to meet energy generation in wind farms made up of variable speed drives. The various components of the VSWT driven type of generator system consisting of its back-to-back voltage source converter with a DC-link and the crowbar protection switch were analyzed. The back-to-back converters consist of a stator or grid-side converter (GSC) and a machine or rotor side converter (RSC). Simulations were performed in PSCAD/EMTDC environment.

The dynamic and transient stability of DFIG VSWT wind farm was investigated. It was concluded that wind farm composed of DFIG VSWT system is more stable during wind speed changes and also grid fault in the power system compared to the use of the fixed speed squirrel cage Induction Generator (IG) wind farm. This is because variable speed drives wind farm can provide sufficient reactive power to stabilize itself during speed changes and grid fault because of the control system of the VSWT topology; hence these drives are very popular and commonly in use in present day for wind turbines. However, despite such characteristics of the fixed speed IG like brushless and rugged construction, low cost, maintenance free and operational simplicity, the transient stability of the IG is very poor because it requires large reactive power to recover the air gap flux when a short circuit fault occurs in the power system. Also, the IG system has limited ability to provide voltage and frequency control, unless enhanced by the installation of power electronic devices and reactive power compensation units like STATCOM, ECS, SMES which increases the overall cost of the wind farm system.

Some key challenges and opportunities in wind farm development like noise, visual effect, wildlife mortality, signal interferences, logistics, government policy and human factors were presented. Some opportunities and way forward to these challenges were mentioned in light of wind farm effective location and development.

Recommendations

Based on the presented simulation results of the VSWT wind farm in this work, it is highly recommended to mix the fixed speed Squirrel Cage Induction generator with VSWT in modern wind farms. This is because the IG cannot be phased out due to the fact that they have some advantages as earlier highlighted. The unstable IG could be stabilized with the DFIG control system, hence wind farms composed of IGs and DFIGs could be promising in the future. Furthermore, it is imperative to consider proper siting of modern wind farms in order reduce some of the key challenges in wind farm development with regards to the environment.

REFERENCES

[1] Babouri R., Aouzellag D., Ghedamsi K. (2013): "Integration of Doubly Fed Induction Generator Entirely Interfaced with Network in a Wind Energy Conversion System," Terra Green 13 International Conference-Advancements in Renewable Energy and Clean Environment, Energy Procedia Science Direct, Vol. 36, pp. 169-178.

[2] Bharat D. Suthar, (2014): "Wind Energy Integration for DFIG Based Wind Turbine Fault Ride Through," *Indian Journal of Applied Research,* Vol. 4, Issue 5, pp. 216-220.

[3] Delarue Ph., A. Bouscayrol A., Tounzi, X. Guillaud and G. Lancigu, (2003): "Modelling, Control and Simulation of an Overall Wind Energy Conversion System," *Renewable Energy*, 28, pp. 1169-1183.

[4] Mohamed Shahbazy, Phillip Poure, Shahrokh Saadete, Mohamed Reza Zolghdri, (2011): "Five Leg Converter Topology for Wind Energy

Conversion System with Doubly Fed Induction Generator," *Renewable Energy,* Vol. 36, pp. 1187-3194.

[5] Ackermann T., and L. Soder, (2002): "An Overview of Wind Energy Status," *Renewable and Sustainable Energy Reviews*, Vol. 6, no. 1-2, pp. 67-127.

[6] Burton T., Sharpe D., Jenkins N., and Bossanyi E., (2001): Wind Energy Handbook, John Wiley and Sons, Ltd.

[7] Badre Bossoufi, Hala Alami Aroussi, ElMostafa Ziani, Ahmed Lagrioui and Aziz Derouich, (2014): "Low Speed Sensorless Control of DFIG Generators Drive for Wind turbines System," *WSEAS Transactions on Systems and Control,* Vol. 9, pp. 514- 525.

[8] Lopez J., P. Sanchis, X. Roboam L. Marroyo, (2007): "Dynamic Behaviour of Doubly Fed Induction Generator during three phase voltage dips," *IEEE Transactions on Energy Conversion*, Vol. 22, no. 3, pp. 709-717.

[9] Mohamed Benbouzid, (2014): "High Order Sliding Mode Control of DFIG-Based Wind turbines," Book Chapter 2, Springer International Publishing Switzerland, pp. 23-48.

[10] Omar Noureldeen, (2012): "Behaviour of DFIG Wind turbines with Crowbar Protection under Short Circuit," *International Journal of Electrical and Computer Sciences IJECS-IJENS,* Vol. 12, no. 3.

[11] Abolhassani M. T., P. Enjeti and H. Toliyat, (2008): "Integrated Doubly Fed Electric Alternator/Active Filter (IDEA), a Viable Power Quality Solution, for Wind Energy Conversion Systems, *IEEE Transactions on Energy Conversion,* Vol. 23, no. 2, pp. ----.

[12] Youcef Bekakra and Djilani Ben Attous, (2011): "Sliding Mode Controls of Active and Reactive Power of a DFIG with MPPT for Variable Speed Wind Energy Conversion," *Australian Journal of Basic and Applied Sciences,* Vol. 5, no. 12, pp. 2274-2286.

[13] Dami M. Ali, Jemli K., Jemli M., Gossa M., "Doubly Fed Induction Generator, with Crowbar System under Micro-Interruptions Fault," (2010): *International Journal on Electrical Engineering and Informatics,* Vol. 2, no. 3, pp. 216-231.

[14] Moulay Tahar Lamchich and Nora Lachguer, Matlab (2012): "Simulink as Simulation Tool for Wind Generation Systems Based on Doubly Fed Induction Machines," MATLAB- A Fundamental Tool for Scientific Computing and Engineering Applications- Vol. 2, Chapter 7, INTECH Publishing, pp. 139-160.

[15] Noubrik A., L. Chrifi-Alaoui, P. Bussy, A. Benchaib (2011): "Analysis and Simulation of a 1.5MVA Doubly Fed Wind Power in Matlab Sim PowerSystems using Crowbar during Power Systems Disturbances, in: IEEE-2011 International Conference on Communications, Computing and Control Applications (CCCA), Hammamet, Tunisia.

[16] Arul I., M. Karthikeyan, N. Krishnan and P. Anush, (2013): "MPPT using Pitch Angle with Various Control Algorithms in Wind Energy Conversion System," *International Journal of Computer Applications,* Vol. 74, no. 7, pp. 15-18.

[17] Bezza M., B. EL. Moussaoui, A Fakkar, T. I. Leea, (2012): "Sensorless MPPT Fuzzy Controller for DFIG Wind Turbine," *Energy Procedia,* Vol. 18, pp. 339-348.

[18] Garcia-Garcia M., M. P. Comech, J. Sallan and A. Liombart, (2008): "Modelling Wind Farms for Grid Disturbances Studies," *Science direct, Renewable Energy*, Vol. 33, pp. 2019-2121.

[19] Kerdoun D., A. Boumassata, N. Cherfia, and N. Bennecib, (2014): "Active and Reactive Power Control of a DFIG with Cycloconverter for Variable Speed WECS," LGEC Research Laboratory, Algeria.

[20] Deok Chul Kim, Joon Ho Choi, Won Wook Jung, Ju Yong Kim and II Keun Song, (2015): "Modelling and MPPT Control in DFIG Based Variable Speed Wind Energy Conversion Systems by using RTDS," *Journal of International Council on Electrical Engineering*, Vol. 1, no. 4, pp. 430-436.

[21] Abdellatif Noubrik, Larbi Chrifi-Alaoui, Pascal Bussy and Abedelkrim Benchaib (2011): "Analysis and Simulation of a Crowbar Protection for DFIG Wind Application during Power Systems Disturbances," *Journal of Mechanics Engineering and Automation,* Vol. 1, pp. 303-312.

[22] Aluko O. T. M. Smith, L. M. Tolbert, (2010): "Behaviour of DFIG under Nearby Wind Plant Fault," IEEE Publishing Company Ltd.

[23] Akhmatov V., (2005): Induction Generators for Wind Power, Multi-Science Publishing Company Ltd.

[24] Salma T., and Yokeeswaran R., (2013): "Pitch Control of DFIG based Wind Energy Conversion System for Maximum Power Point Tracking," *International Journal of Advanced Research in Electrical, Electronics and Instrumentation Engineering*, Vol. 2, Issue 12, pp. 6373-6380.

[25] Holdsworth L., X. G. Wu, J. B. Ekanayake, N. Jenkins, (2003): "Comparison of Fixed Speed and Doubly Fed Induction Wind turbines during Power System Disturbances, in: IEE Proceedings- Generation, Transmission and Distribution, May 13, Vol. 150, Issue 3, pp. 343-352.

[26] Francois B., Yongdong Li, (2009): "Improved Crowbar Control Strategy of DFIG based Wind turbines for Grid Fault Ride Through," Applied Power Electronics Conference and Exposition, pp. 1932-1938.

[27] Peng L., (2010): Reconfiguration du dispositife de Commande d'une ecolienme en cas de Creux de Tension, Thesis, L2EP Lille.

[28] Slootweg J. G., H. Polinder, W. L. Kling, (2001): "Dynamic Modelling of a Wind Turbine with Doubly Fed Induction Generator," in Proceeding of IEEE Power Engineering Society Summer Meeting, Vol. 1, pp. 644-649.

[29] Koessler R., S. Pillutla, L. Trinh, D. Dickmander, (2003): "Integration of Large Wind Farms into Utility Grids: Part 1, Modeling of DFIG," in Proceeding of IEEE PES General Meeting.

[30] Bollen M. H. J., M. Martins, G. Olguin (2004): "Voltage Dips at the Terminals of Wind Power Installations," in Nordic Wind Power Conference, Chalmers University of Technology, Goteborg, Sweden.

[31] Runcos F., R. Carlson, A. M. Oliveir, P. K. Peng, N. Sadowski, (2004): "Performance Analysis of a Brushless Double Fed Cage Induction Generator," in Nordic Wind Power Conference 2004, Chalmers University of Technology, Goteborg, Sweden.

[32] Niiranen J., (2004): "Voltage Dip Ride Through of Doubly Fed Generator Equipped with Active Crowbar," in Nordic Wind Power Conference, Chalmers University of Technology, Goteborg, Sweden.

[33] Muller S., M. Diecke, W. Rik., and De Doncker, (2002): "Doubly Fed Induction Generator Systems for Wind turbines," *IEEE Industry Applications Magazine,* pp. 26-33.

[34] Slemon G. R., (1989): "Modeling of Induction Machines for Electric Drives," *IEEE Trans. Ind. Application*, Vol. 25, no 6, pp. 1126-1131, Nov. /Dec.

[35] Karim Belmokhtar, Mamadou Lamine Doumbia and Kodjo Agbossou, (2011): "Modelling and Power Control of Wind Turbine Driving DFIG Connected to the Utility Grid," International Conference on Renewable Energies and Power Quality, Spain.

[36] Petersson A., (2005): "Analysis, Modeling and Control of Doubly-Fed induction Generators for Wind turbines," PhD Thesis, Division of Electric power Engineering, Department of Energy and Environment, Chalmers University of Technology, Sweden.

[37] Godoy M. Simoes and Felix A. Farret, (2009): "Renewable Energy Systems Design and Analysis with Induction Generators," Power Electronics and Application Series, University of West Florida.

[38] Min Min Kyaw, V. K. Ramachanaramurthy, (2011): "Fault Ride Through and Voltage Regulation for Grid Connected Wind Turbine," *Science Direct, Renewable Energy*, Vol. 36, pp. 206-215.

[39] K. E. Okedu, S. M. Muyeen, R. Takahashi, and J. Tamura, (2012): "Wind Farms Fault Ride Through using DFIG with New Protection Scheme," *IEEE Transactions on Sustainable Energy*, vol. 3 no. 2, pp. 242-254, April.

[40] "PSCAD/EMTDC Manual" (1994), Manitoba HVDC research center.

[41] K. E. Okedu, (2010): "The Effect of a Resistance Magnitude of a Crowbar Switch on the Protection of a DFIG during Grid Fault," *International Journal of Applied Engineering and Research IJAER*, vol. 5, no.14, pp. 2489-2494.

[42] K. E. Okedu, (2012): "Stability Enhancement of DFIG-based Variable Speed Wind Turbine with a Crowbar by FACTS Device as Per Grid Requirement," *International Journal of Renewable Energy Research*, vol. 2, no. 3, pp. 431-439 (available online).

[43] K. E. Okedu, (2010): "Wind Park Stabilization with External Compensation Device," *International Journal of Applied Engineering and Research IJAER*, vol. 5, no. 12, pp. 2173-2181, (available online).

[44] K. E. Okedu, S. M. Muyeen, Rion Takahashi, and Junji Tamura (2011): "Participation of FACTS in Stabilizing DFIG with Crowbar during Grid Fault Based on Grid Codes," 6th IEEE-GCC Conference and Exhibition, Dubai, UAE, pp. 365-368, February, 2011, Paper Number 1569339245 (available online ieee xplorer).

[45] K. E. Okedu, S. M. Muyeen, R. Takahashi, and J. Tamura, (2011): "Comparative Study on Current and Voltage Controlled Voltage Source Converter based Variable Speed Wind Generator," Proceedings of International Conference on Electric Power and Energy Conversion Systems (EPECS`2011), Sharjah, UAE, Nov. 15-17, 2011 (available online ieee xplorer).

[46] K. E. Okedu, S. M. Muyeen, R. Takahashi, and J. Tamura, (2012): "Wind Farm Stabilization by using DFIG with Current Controlled Voltage Source Converters Taking Grid Codes into Consideration," *IEEJ Transactions on Power and Energy*, vol. 132, no. 3. pp. 251-259 (available online ieej xplorer).

In: Wind Farms
Editor: Marian Dunn
ISBN: 978-1-63484-841-1

Chapter 2

Assessing Noise from Wind Farms

Valeri V. Lenchine [*] ***and Jonathan Song***
Science, Assessment and Planning Division,
SA Environment Protection Authority, Adelaide, Australia

Abstract

Wind farms have demonstrated impressive growth in electricity generation capacity over the past decades. Alongside this growth trend, some communities living in areas adjacent or close to existing and future wind farm sites have expressed concerns regarding possible health and environmental implications resulting from wind farm operations. Among the environmental concerns of wind turbine operations is the noise impact from wind farms.

A wind farm operation should meet certain requirements in terms of noise impact. These noise limits are normally imposed by regulatory or planning authorities and are typically one of the strictest limits to be applied to potential noise sources. In many cases noise from wind farm operations is just above background or ambient noise present. Therefore monitoring and compliance checking of wind farm operation noise may be a complex scientific and engineering task.

This chapter explores a variety of methods to be used for assessing noise from wind farms. The advantages and shortcomings of each approach to wind farm monitoring are discussed and considered within

[*] Corresponding author: email: Valeri.lenchine@sa.gov.au.

this chapter. Recommendations on implementations are provided based off the practicability and accuracy of results produced.

INTRODUCTION

Electricity supply from Renewable energy sources is growing at impressive rates around the world. A substantial part of the green energy demand is met by large scale wind farm developments. However, environmental concerns, whether legitimate or not, may create serious obstacles for the expansion of renewable energy.

The commercial viability of wind farm developments is based on the ability of the wind farm to provide a reasonable power output 24 hours a day, 7 days a week, or for at least a substantial fraction of a year. A largely known side effect of wind farm operations is noise emission. Noise may impact residents living in areas adjacent to operational wind farms. Recognizing the almost permanent presence of wind farm noise for the residents, development approval conditions or other regulatory tools frequently impose very strict rules with respect to noise from wind farms. As a consequence of the low noise limits, it may be difficult to extract noise contribution from wind farms using data measured at a distant receiver. Therefore, the analysis of wind farm noise may represent a challenging scientific task where results are frequently not conclusive.

This chapter discusses the typical approaches implemented to limit noise exposure from wind farms. A variety of methods used for assessing noise from wind farms is considered in the paper. The advantages and shortcomings of each approach to wind farm noise monitoring are discussed and considered within this chapter. Recommendations on implementations are provided based off the practicability and accuracy of produced results.

1. WIND FARM NOISE: PECULIARITIES OF IMPACT

One of the most problematic aspects of wind turbine generators (WTGs) is noise present in areas adjacent to the wind farms. Industrial onshore wind farms require a significantly large area to deploy tens and sometimes hundreds of high capacity wind turbine generators. Therefore, wind farms are normally planned and designed to be placed in rural areas. Background or ambient noise levels in these zones are typically much lower when compared with metropolitan areas. Residents of rural areas commonly expect a high level of environmental amenity. Many of the residents affected by wind farm noise believe that wind farms should be totally inaudible. However, in many countries the impact from a wind farm is considered acceptable if the noise levels do not exceed certain limits. This is consistent with approaches applicable to other noise sources.

It should be noted that meeting the expectations of every resident living in the affected zone is almost impossible. There are too many factors influencing the perception of wind farm noise by an individual listener such as visual impact, income stream from the development, higher sensitivity, etc. Regulatory limits are frequently based on population noise exposure research. For example, European Guide on Noise Exposure shows that even if such a strict criteria as 45 dB(A) (day-evening-night descriptor L_{den}) is met, about 6% of population is still expected to be annoyed [1].

Apart from acceptability of the noise for majority of the affected population, rationales of achievability and practicability are taken into account. *Night Noise Guidelines for Europe* [2] recommend a 30 dB(A) outdoor level as an ultimate goal from a health perspective. However, the Guidelines recognize that this goal is not realistic at the moment and sets an interim goal for night time noise levels of 55 dB(A) as more feasible and achievable. Similar to that, approaches to controlling noise from wind farms may take into account a number of factors to set reasonable and practicable limits. The protection of amenity of affected residents is important, but striking a balance between the amenity and energy generating developments is also vital for the progress of a modern society.

1.1. Typical Noise Limits

The specification of noise limits for environmental noise sources can be based on the expected level of amenity for relevant receivers (baseline noise criteria) or the typical background noise present in the area. The latter relies on the masking effect of background noise which makes noise from the source of interest less intrusive or annoying. There is a variety of approaches for applicability of the noise criteria. Relevant or noise sensitive receivers normally include dwellings where people live permanently but do not have commercial interest in the wind farm development. It may also include hotels, caravan parks, hospitals, etc. There may be more relaxed noise criteria for owners of houses who have an income stream from hosting the wind turbines on their property or participating in the commercial development.

Modern wind farms may have a very large environmental footprint and affect houses located a few kilometers away from the wind farm. Wind farm noise at large separation distances is influenced by many factors including but not limited to variations in the wind speed and direction, atmospheric temperature distribution, presence of turbulence structures, etc. It is expected that fluctuations of wind farm noise measured at a distant receiver may be significant. Therefore the specified criteria are to be met statistically, which means that measured wind turbine noise can exceed the noise criteria at times. Equivalent A-weighted sound pressure levels (SPLs), or metrics based on that, is the most widely used descriptor for wind farm assessment. Allowable noise limits are typically specified as outdoor levels. It is impracticable and in some cases impossible to predict indoor levels for each of room located in noise affected dwellings.

The sound power of a wind turbine is correlated and reported versus the turbine hub height wind speed [3]. This assumption of a direct link between the hub height wind speed and noise from the wind farm is critical for specifying noise limits from a wind farm. It should be noted that older versions of relevant standards, guidelines or assessment procedures utilized wind speed data collected at 10 m above ground. This height is consistent with meteorological standards and was an obvious choice for the assessment of wind turbine noise. However, the 10 m reference was not always a good indicator to characterize a wind turbine as a noise source. For example, wind speed at the hub height cannot always be calculated accurately from 10 m indications due to variations in the site wind shear. Consequently, estimation of the noise emission from a wind turbine based on wind speed at 10 m above the ground may bring significant error in the prediction and assessment of noise from a wind farm. To eliminate this impediment, modern methods of wind farm assessment specify hub height or similar height on erected masts as a reference for prediction and measurement of the noise.

Alternative or combined noise criteria are common for limiting noise impact from wind farms. The latter allows for choice between a baseline noise criteria or "background + tolerance" noise criteria for particular wind speeds where the greater limit is to be chosen. For this paragraph, baseline criterion is considered first. This limit is linked to the expected level of amenity where the affected dwellings are located. Actual limit can be established by the planning authority, either State or local governments. The criteria differ significantly from country to country or even state to state. Specifications may also be different for day and night time limits with the latter sometimes being significantly stricter. Analysis of available information indicates that the range of baseline noise criteria for wind farms vary from 35 to 70 dB(A) [4, 6]. The strictest noise criteria are normally attributed to rural living areas with a high level of amenity in sparsely populated zones and the absence of agricultural production activities.

Many European countries utilize day-evening-night SPL (L_{den} descriptor) which includes penalties added to the measured evening and night time levels. If a noise sensitive receiver is exposed to 35 dB (A) throughout 24 hours, it will be roughly equivalent to 41 dB (A) of L_{den}. This number is marginally above the second interim goal of 40 dB (A) for night time noise in Europe and comfortably bellow the first interim goal of 55 dB (A) [1].

The other way of specifying noise criteria is based on background noise measured prior to construction of the wind farm. The environmental impact from a wind farm may be considered acceptable if its operation only marginally increases noise levels at the relevant receiver:

$$L_{WF} = L_B + \Delta L,$$

where L_{WF} is the permissible wind farm noise, L_B is the pre-construction background level and ΔL is the allowable tolerance. The most common permissible increase above the background level is 5 dB(A) which is often indicated as the first clearly perceivable difference. There is precedence to reduce the tolerance to 3 dB(A) for night time hours [5] which is classified as just perceivable.

As noted above, significant variations in environmental noise levels are not rare. Similar to meeting noise limits from a statistical perspective, the background noise is understood as an average indicator for a particular wind speed measured at the proposed wind farm site. Sometimes L_B can be derived separately for night time and day time. Such variation in noise limit specification may be justified if the background noise differs significantly between day and night time hours.

Derivation of the statistical background curve versus wind farm wind speed may be based on different data post-process techniques. The acoustic descriptor L_{A90} reported for 10 minute intervals is commonly used for environmental noise monitoring to represent background noise levels. This descriptor is equivalent to the A-weighted SPL present in the environment for

90% of the time. It is used as a statistical filter to reduce the influence of ambient and extraneous noise sources. A typical requirement for environmental noise monitoring such as rectification of the data set from data affected by rains, high local wind speeds and known extraneous noise events are applicable for wind farms noise monitoring. Nevertheless, variations in the background levels may be significant. The background SPLs can be better established if data post-processing:

- limits the data points to that collected between the cut-in wind speed (when a turbine starts the generation of electricity) and the speed of the rated power (corresponds to the maximum of electricity generation);
- limits the polynomial order of the statistical curve fitting (typically polynomial order should not be above 3);
- uses a sufficient number of data points for each of the wind speed ranges.

More details on background data acquisition can be found in Section 3.

Figure 1 shows an example of a background noise analysis based on a data post-process which utilized the principles above. It shows a combined noise criteria comprising a baseline criterion of 35 dB(A) which is then replaced by a "background + 5dB(A)" limit when it becomes greater than the baseline limit at around 12 m/s wind speed. The involvement of data collected outside of the recommended range of the wind speeds sometimes leads to a significant distortion of the background curve and unrealistic trends. The rationale behind exclusion of the data above the speed of the rated power (in some procedures 95% of the rated power [7]) is the assumption that since the turbine does not produce more electricity at the higher wind speeds it also does not emit more noise which is efficiently masked by background at high wind speeds. The inclusion of the points above the rated speed into the data set may be detrimental to the resulting background curve. Some methods of background assessment recommend further segregation of data for particular wind directions [7] or the worst case downwind scenario is considered more appropriate for noise limit specification and consequent compliance checking [8].

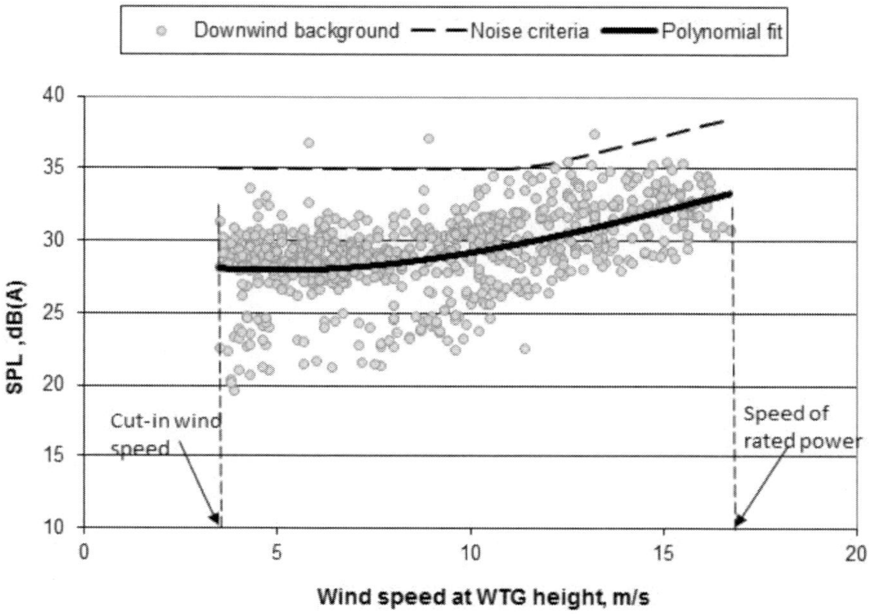

Figure 1. Combined noise criteria derived from downwind background (L_{A90} descriptor).

1.2. Noise Characters (Tones, Amplitude Modulation, Low Frequency Noise)

The presence of noise characters in a wind turbine noise emission can exacerbate the perception of a listener. People living next to wind farms can sometimes find wind farm noise particularly annoying due to noise characters even if the overall noise levels are low. The operation of a wind turbine may be accompanied by one or more of these noise characters:

- Low frequency content;
- Tone generation;
- Amplitude modulation.

Assessment of any of the characters at a distant receiver may represent a difficult task due to very low overall noise impact and the absence of accurate methods of their assessment. Therefore, some of the relevant noise assessment procedures address this by assigning strict noise limits, recognizing the fact that some of the characters may sometimes be audible at noise sensitive receivers [8]. The presence of a character sometimes involves the application of a penalty to the measured A- weighted levels, which may be crucial for formal compliance with relevant regulatory limits.

Noise is characterized as having a low frequency character if it has significant spectral components below 200-250 Hz [7, 10]. The lower boundary of the frequency span tends to extend down to 8-10 Hz in accordance with some of relevant methods of assessment [11, 12]. This is considered to be well into the infrasonic region. Even if the sound power characteristics of a wind turbine does not raise concerns about low frequency noise, the character may be a potential problem at a distant receiver. This is due to higher attenuation for mid and higher frequencies while propagating over significant separation distances or through a dwelling structure when compared to the lower frequency noise components.

Since low frequency noise is a wider problem which can be attributed to common environmental and equipment noise sources, there is a number of approaches that can be employed for evaluating low frequency noise as applicable to wind farms. One group of acoustic metric is based on C-weighted SPL or the arithmetic difference between C-weighted and A-weighted noise levels. If the difference between the levels acquired with the different weightings exceeds 15-20 dB (or 25 dB for very low overall noise levels) [10], it is indicative of a potential for the low frequency problem. Study [13] suggests a simple method of assessing low frequency impact based on C-weighted noise levels. The author points out that outdoor levels of 60-65 dB(C) during night time hours are sufficiently conservative to preserve amenity expected in residential areas. More complex methods for low frequency character assessment include analysis of third octave levels (UK Department of Environment, Food & Rural Affairs), low frequency criteria for 10-160 Hz 1/3 octave central frequencies [11] or the introduction of special acoustic descriptors. For example, the Danish Environment Protection Agency introduced a low frequency acoustics descriptor to be used for the assessment of noise impact for new developments. $L_{pA,LF}$ represents A-weighted SPL calculated for 1/3 octave span from 10 Hz through 160 Hz. Danish *Statutory Order on Noise from Wind Turbines* [14] recommends use of the descriptor for a limited range of environmental conditions to predict wind turbine noise inside of affected dwellings. Level $L_{pA,LF}$ of 20 dB(A) for evening and night time hours is considered appropriate to ensure an absence of the low frequency problem.

All of the criteria and methods above may be difficult to apply in practical cases. Natural background and wind induced noise affecting the measurement microphone and other noise sources may frequently generate similar or even higher low frequency noise levels. In the case of indoor noise measurements, operation of home appliances, pumps and human activities make the task of separating wind farm contribution from other noise sources more complicated. These predicaments may explain the absence of a widely accepted consensus regarding methods of objective assessment for wind farm low frequency noise and penalties applied to the measured noise levels if a breach of the recommended low frequency criteria is confirmed or predicted.

A recent long term study of wind farm noise was funded by Japanese Ministry of Environment. It involved a significant amount of noise data measured in the vicinity of 34 wind farms. The study confirmed that A-weighted SPL is an adequate primary metric for the analysis of wind farm noise. The study also indicated a high degree of correlation between C-weighted and A-weighted levels in a wind turbine controlled environment [16] which may eliminate the need for a separate assessment of C-weighted levels. The available information shows that if noise from a wind turbine meets a 45 dB(A) criterion, it will also comfortably meet the C-weighted low frequency noise criterion suggested in [13].

A wind turbine may potentially emit tones which are typically connected to the operation of mechanical components within the turbine nacelle. Standard [3] details the procedure for reporting tones in a wind turbine controlled environment. If audible at a distant receiver, the character may evoke the applicability of a 5-6 dB(A) penalty to the measured SPL [7, 8]. The procedure of tonality assessment in IEC Standard [3] is primarily intended for measurements performed close to the turbine. It may not be directly applicable to the noise measurements at a distant receiver. The standard procedure of the tonality assessment requires correction of measurements for the background noise. It would be reasonable to consider existing background as part of the masking noise at a distant location to decide on audibility of the tone in an environment where noise is not always controlled by the wind turbines. In that case, the IEC method can be modified or the procedures for tonality assessment of a general environmental noise can be employed.

Another characteristic of a wind turbine emission is an amplitude modulation which is always present at close proximity to a wind turbine and to some degree may be audible over significant separation distances. This effect is frequently characterized as a “swoosh” sound and represents a periodic variation of the noise level. Amplitude modulation is frequently confused with impulsiveness. Impulsive sound, such as hitting a sheet of steel by a hammer, is characterized by an abrupt raise and fall of the noise levels and a characteristic time below 100 milliseconds [7]. Impulsive noise character is typically not a characteristic of a well maintained and operated turbine.

Listeners can discern a very small modulation with modulation depth of just 1 dB [17]. Some of the methods of assessment consider the modulation depth in a generic sense, thus replacing its estimate by the peak-to-trough difference. The latter approach is applicable to analysis of SPL time traces with clear amplitude modulation pattern. Some modifications of the method are suggested in study [16] which involves the analysis of the difference between SPLs acquired with "Fast" and "Slow" time weightings. If the modulation pattern is not explicit and measurements are affected by background or extraneous noise, it calls for more complex methods of analysis [17], [18].

Excessive amplitude of modulation is considered as a factor exacerbating the annoyance of a listener. Therefore, excessive modulation may be penalized under different noise assessment procedures. A 5 dB(A) penalty applicable to the measured A-weighted SPL is suggested in the New Zealand Standard [7] if the peak-to-trough variations consistently exceed 5 dB for 2 minutes or more during a 10 minutes integration period. However, a comprehensive study commissioned by RenewablesUK [17] does not recommend any particular system of penalties for excessive modulations. The study indicates that a modulated sound with parameters typical for wind farm noise is slightly more annoying than non-modulated sound, but does not recommend any particular threshold when the modulation can be considered excessive and evokes a penalty.

It should be noted that the noise characters considered in this section represent a number of additional effects that may exacerbate perception of noise from a wind farm. In general, the combination of noise emission of a particular turbine and atmospheric conditions may cause different sensations to a listener. For example, wind turbine noise with a high degree of amplitude modulation is frequently characterized as "throbbing" or "thumping." Sensitive listeners reported a "rumbling" sound from a wind farm, which one would expect from the operation of air conditioning equipment [19]. Such effects are not typically observed at many receivers and not for a significant fraction of the time.

1.3. Note on Infrasound Impact

Some earlier wind turbines were based on designs with downwind positioning of the blades from the turbine tower. Aerodynamic "shade" from the tower caused a significant increase of amplitude modulation and likely infrasound emission as well. Modern industrial turbines are controlled with the blades kept upstream from the tower. Sensitive people residing in the vicinity of a wind farm often attribute excessive infrasound from the wind farm as a reason for implied adverse health effects experienced.

International standard ISO 7196 [20] for infrasonic measurements did not establish explicit acceptable levels for infrasound. Nonetheless, the standard states "Weighted sound pressure levels which fall below about 90 dB (G-weighting is meant, *auth.*) will not normally be significant for human perception." A number of researchers suggest a conservative hearing threshold for infrasound of 85 dB(G) when measured in accordance with the methodology in the standard [21]. This threshold accounts for possible variations in individual sensitivity of listeners to tones and broad band infrasound. Medical research dedicated to infrasound finds that there are no proven physiological effects at infrasound levels below the perception threshold [22, 23, 24].

Modern wind turbines do not generate infrasound at levels above 85 dB(G) [3]. A listener has a greater risk of exposure to higher infrasound levels in an urban environment rather than in the vicinity of a wind farm [25]. Therefore, the authors are not aware about any modern infrasound assessment procedures and recommended or mandatory infrasound limits that form a part of the development approval or compliance checking procedure as applicable to wind farms.

2. Approaches to Assessment of Noise in Wind Farms

From the information given in the previous sections, it is clear that there may be a range of approaches to assessing the environmental impact for new and existing wind farms. They may reflect peculiarities of local planning and environmental assessment policies and relevant legislative frameworks. Procedures applicable to the noise assessment of wind farms are periodically updated to reflect state of the art relevant knowledge or changes in available acoustic instrumentation and data analysis methods.

2.1. Development Assessment and Post-Construction Stages

Modern industrial type wind farm developments may be comprised of hundreds of wind turbines, each with the capacity to generate a few megawatts of electricity. The developments cover significant areas and zones. A wind farms' environmental footprint extends far beyond geographical boundary of the site. Hub height of conventional wind turbines can reach 100 m above ground or even greater. Installation of turbines on hills and elevated areas generally improves conditions for harvesting energy from wind. At the same time, this increases the elevation of noise sources and the area affected by wind turbine noise. Since the potential for noise impact of a wind farm is substantial, development approval authorities require a noise assessment be made as part of the Environmental Impact Assessment Statement or a similar document. Compliance with applicable noise limits should be proven prior to attaining approval for a wind farm development. Post-construction noise monitoring should confirm compliance with the applicable limits or indicate that measures should be undertaken to reduce noise impact in the case of non-compliance.

In the majority of cases it is recommended to perform pre-construction background noise measurements at relevant receivers. The data can be used for:

- establishing noise criteria based on the "background + tolerance" rule;
- utilizing the data for the compliance checking procedure if correction for the background is expected to be necessary.

Even if the prediction of noise impact indicates that the wind farm development will meet requirements of applicable baseline criterion and there is no formal requirement to measure the background noise, the background data can be useful in the future. The predicted impact may be inaccurate and the availability of background data helps in setting alternative noise criteria that may be crucial in resolving the non-compliance situation. The background data will also be of use if the correction for background procedure is necessary to prove the acceptability of the noise impact. Collection of background data after commissioning of a wind farm may be an extremely expensive exercise, especially if shutdown of the entire site is required for the collection of sufficient amount of data points for a variety of environmental conditions.

2.2. Outdoor Noise versus Indoor Noise

The most widely used practice for the assessment of environmental impact of wind farms is the prediction of outdoor noise levels at a relevant receiver. Indoor noise levels are often affected by individual room acoustics and design of the houses. Moreover, compliance checking measurements of noise inside of an occupied dwelling is normally significantly affected by household activities, operation of home appliances and tools [25].

One possible way of facilitating the prediction of noise impact inside houses is the concept of a "standard" house where structure of the houses is considered to be providing statistically average transmission loss. This concept was implemented for predicting wind farm noise indoors in some European countries [14]. However it did not attain wide acceptance around the world due to too many factors which challenge the value of such an approach.

2.3. Acoustic Descriptors Used

Regulatory requirements for wind farm developments are strict in developed countries. The typical allowable noise limit for residential noise receivers are between 35-50 dB(A). The sensitivity of a human ear to small levels of noise is approximated by A-weighted noise levels which explains the fact that equivalent A-weighted SPL or metrics based on its magnitudes is the dominant acoustic descriptor applied to wind farm noise assessment.

There may be different groups of descriptors which are utilized for specification characterizing a wind turbine as a noise source. In addition to A-weighted sound power of a wind turbine, the range of acoustic descriptors of a wind turbine model may include 1/3 octave A-weighted levels for a range of operating wind speeds and the results of tonality assessment based on A-weighted spectral analysis (narrow band) [3]. A comprehensive acoustic specification of a wind turbine provides an informed input for the prediction of noise impact from a wind farm and should be obtained from the turbine manufacturer at the development application stage. This is important for securing accuracy of the noise impact modeling.

Measurement procedures at a distant receiver are typically based on statistical descriptors derived from A-weighted SPL such as 90- or 95-percentiles, shorten to L_{A90} or L_{A95} respectively. Analysis of C-weighted or 1/3 octave data is normally not required unless problems with noise characters are reported. As it was noted in the previous section, contribution of a wind farm at a distant receiver may be insignificant and the calculation of a wind farm impact in an environment where the turbines do not control noise represents a challenging technical task. This is particularly true if unattended monitoring of wind farm noise should be performed for a prolonged period using conventional acoustic instruments.

3. Conventional Methods Used for the Assessment of Wind Farm Noise

3.1. Assumptions for Assessing Noise from Wind Farms

Assessing noise from wind farms considered to be a challenging technical task. As noise assessment may take place at residential locations a few kilometers away from the nearest wind turbine, actual noise contribution from the wind farm could be just above the background noise level or may only be audible at particular times. Another essential factor causing complications in this task is wind induced noise as wind farms are situated in areas with high wind speeds. High wind speeds cause significant interference when it comes to noise measurements. In order to counter this issue, sophisticated wind shields or advanced signal processing techniques [26] are required to be used to account for the impact from wind noise.

A few assumptions are required to be made when assessing environmental impact from wind farms. It is assumed that noise from wind farms is correlated to hub height wind speeds of the wind turbines. Noise from a wind turbine at a distant receiver is expected to be consistent with the sound power characteristic of the wind turbine. It is also assumed that sound power of a wind turbine does not increase after reaching the speed of the rated power. Background noise at higher wind speeds (above the speed of the rated power) is considered to be sufficient to mask wind turbine noise. When measuring noise from wind farms, noise below the cut-in speed (when wind turbine starts generating electricity) can be assumed to be not associated with the wind farm; while noise above the rated speed of the wind turbine is supposed to be mostly controlled by background noise. Therefore, noise impact prediction and assessment outside of the wind speed range is not required.

If there are many noise sensitive receivers affected by the operation of the wind farm, the concept of a "representative location" can be called on for the noise monitoring. It is assumed that the noise impact and environmental condition at the "representative location" is similar to that of the neighborhood houses or a group of the houses. This allows for noise and weather data to be acquired at one location rather than at several similarly positioned receivers.

3.2. Requirements for Wind Farm Noise Assessment Procedures

Modern acoustic instruments and data processing techniques allow for the development of methods to extract the contribution of a wind farm in the total measured noise. However, the implementation of complex methods frequently involves expensive instruments, complicated data post-processing and advanced knowledge of acoustics and data analysis.

Preferably, wind farm monitoring procedures should be based on:

- commercially available and inexpensive instruments such as noise loggers or sound level meters;
- conventional acoustic descriptors which are readily available;
- relatively simple and reliable methods of data analysis which bring repeatable and consistent results.

Noise monitoring procedures that meet the requirements above can be considered practicable and may form a part of a relevant standard or regulatory document.

The noise data collected should be synchronized with the wind farm wind speed measurements. Averaging periods of 10 minutes is normally utilized for commercial wind farms. Data acquisition systems should meet the requirements specified in the relevant standard or procedure. The majority of the documents recommend that the measurement system should meet the requirements of the Class 1 noise level meter specified in IEC-61672.1 standard [27]. Noise floor of the acoustic instruments should be lower than typical background levels at the monitoring location to allow for accurate acquisition of the acoustic parameters. Low levels of noise are not rare in areas adjacent to a wind farm and it may require that commercially available instruments with the lowest self-noise be used for data acquisition. The monitoring instruments should be able to report a set of acoustic descriptors and weather parameters (wind speed and direction, precipitation, temperature, etc.) utilized by a particular noise monitoring method.

It is difficult to indicate methods for noise character assessment which could be generally acceptable for wind farm noise monitoring at a distant receiver. Therefore, only methods based on total noise measurements are considered in this section.

3.3. Typical Noise Assessment Procedures

In this section, the methods of noise assessment considered are based on the acquisition of total noise by acoustic instruments such as noise loggers and sound level meters. These methods are most widely used for assessing environmental impact from wind farms in spite of the fact that the mathematical justification of the methods may not be of the highest standard. The relative simplicity and practicability of the procedures have secured their use for decades.

3.3.1. Pre- and Post-Construction Noise Levels

As it was noted earlier in the text, noise criteria may be based on the "background + tolerance" rule. This requires background data acquisition to be made prior to the construction of a wind farm.

Data acquisition instruments should be positioned on the side of a noise affected dwelling facing the wind farm. The position of the instrument should not be screened from the wind farm by the house or other associated structures. The influence of other noise sources such as vegetation, pumps, air conditioners, etc. should be minimized. The microphone should be placed 1.2-1.5 m above the ground and at least 3.5 – 5 m away from vertical reflecting surfaces. However, local requirements for the measurement of wind farm noise may differ.

Since wind induced noise may be a potential problem for data acquisition, the microphone should be equipped with a wind shield and local wind speed measurements at microphone height should be performed to monitor the wind speed against the wind shield performance. Ultrasonic weather sensors are preferable for the measurements since they do not contain moving parts such as weather vanes or cup sensors and therefore do not contribute significantly to the measured noise.

Collected acoustic descriptors should be synchronized with the indication of a reference wind speed. Wind speed and direction should be measured at the nearest meteorological mast installed on a future wind farm site. The wind speed data should be acquired at the turbine hub height or at other heights which enables accurate calculation of the reference wind speed [3]. Some European procedures based on an older approach where instead of the hub height reference wind speed the meteorological height of 10 meters above the ground is used. Many commercial wind farms are equipped with SCADA data reporting system which provides wind data for 10 minutes intervals. Modern procedures use L_{A90} descriptor (equivalent A-weighted SPL equaled to or exceeded for 90% of the period). This serves as a statistical filter to minimize the influence of extraneous or ambient noise on the measurements.

The collection of data should provide statistically sufficient amount of data to derive the background curve or background data corresponding to particular wind speeds. The data array should have sufficient amount of data points for the wind speeds of interest and particular wind directions. The requirements for the number of points may vary but at least 1000 - 2000 valid data points during a monitoring period is suggested. Some methods are more concentrated on the assessment of worst case scenario and require a certain number of data points to be collected for the downwind direction [8]. In accordance with ISO 1996:2, downwind is considered to be the wind directions within ±45° from a direct line connecting the noise source and receiver.

The first step to data analysis is the elimination of data affected by extraneous noise sources, rain and high wind induced noise. Some of the procedures recommend disregarding data gathered at local wind speeds above 5 m/s unless there is sufficient technical evidence that the wind shield used was able to provide good performance at high wind speeds [8]. The next step taken is analyzing the available array of valid background data. There are two conventional approaches to the data analysis:

- curve fitting;
- data post- process in the wind speed bins.

Polynomial fit should be based on all valid data between the cut-in and speed of the rated power of the wind turbine. This method is better for analysis of the trend versus the wind speed. Polynomial (or linear) curve should be chosen based on the best fit providing a higher coefficient of determination and trend consistent with the wind speed change. High polynomial orders frequently bring unrealistic waviness to the trendline, therefore the curve fit should be limited to the second or third polynomial order.

Post-process for wind speed bins is based on segregation of data into K wind speed bins. If the data is supposed to be reported for integer wind speeds, the boundaries of the wind speed bins should be chosen as follows:

$$V_k - 0.5 \leq V < V_k + 0.5,$$

where V_k is an integer wind speed in m/s.

The background SPL in the wind speed bin is calculated as the energy average of all data points within the wind speed bin:

$$L_k = 10 \text{ Log}\left(\frac{1}{M}\sum_{i=1}^{M} 10^{\frac{L_i}{10}}\right),$$

where L_i is the SPL for i-th measurement and M is the number of background measurement points in the wind speed bin.

The advantage of this type of post-processing is that it is less susceptible to data which is not representative of the wind farm noise if they are distributed over the entire range of wind speeds. On the other hand, if many poorly rectified data points are concentrated in a particular wind speed bin, it can give an erroneous estimate of background SPL which is noticeably different from the neighboring magnitudes and violates the general trend.

In the majority of cases, both of the methods described give consistent results as seen in Error! Reference source not found.

Compliance checking methods are utilized to provide assessment of a wind farm during or after commissioning or to investigate grounds for complaints of the residents. The procedure is similar to background noise monitoring. Let us consider some features that may be different for the post-construction data acquisition.

The temporary weather mast may be dismantled after wind farm commissioning, so the same wind speed reference may not be available for post-construction noise monitoring. The reference can be replaced by meteorological stations from a similarly positioned permanent weather mast or hub sensor from the nearest turbine. In general, a wind farm layout may be complex and two or even more groups of wind turbines can be separated by a similar buffer from noise sensitive receivers. Data analysis for multiple wind direction references is expected for these cases.

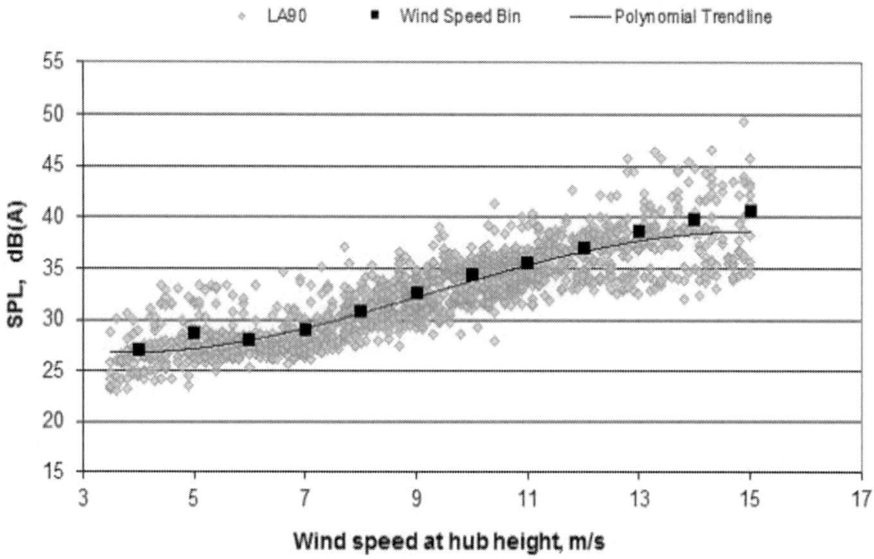

Figure 2. Polynomial fit and wind speed bin post- processing of background noise levels (L_{A90} descriptor).

Some of the compliance checking procedures require the addition of correction to measured wind farm noise to accommodate for the difference between L_{Aeq} and L_{A90} descriptors. This correction is expected to be in the order of 2-2.5 dB in wind farm controlled noise [16].

Variations in the background and wind farm noise assessment procedures include separation of data depending on night and day time periods, consideration of the data for particular wind directions or downwind only. If the total post- construction noise meets applicable noise limits, further investigation of wind farm noise is not required. Otherwise, correction for background should be applied to extract the wind farm contribution. This method is considered in the next section.

3.3.2. Correction for Background Method

In situations when overall noise data acquired after commissioning of a wind farm does not bring conclusive results for development approval requirements or complaint investigations, correction for background procedure is expected to be employed for further analysis. Similar to what is described in the previous section, the procedure may involve either curve fitting or data analysis of the wind speed bins. In the first case, the curve fitting is performed for total noise and the relevant background data. To derive wind farm noise, operation of the energy subtraction is performed for a wind speed of interest.

The wind speed bin procedure involves the calculation of the energy average levels in a particular wind speed bin for total noise and background data and then the energy subtraction procedure is applied to these averages:

$$L_k = 10\ Log(10^{L_{Tk}/10} - 10^{L_{Bk}/10}),$$

where L_{Tk} is the total noise and L_{Bk} is the background noise for k-th wind speed bin.

As it is noted before, some of the noise assessment procedures presume that the wind farm noise calculations should be made under particular environmental conditions such as during downwind periods.

Figure 3 shows an example of correction for the background method based on the total noise and background curves energy subtraction. The picture indicates that this procedure brings noticeable reduction to the estimated wind farm noise in comparison with the overall noise measurements. It may be essential for proving compliance of a wind farm noise with applicable limits.

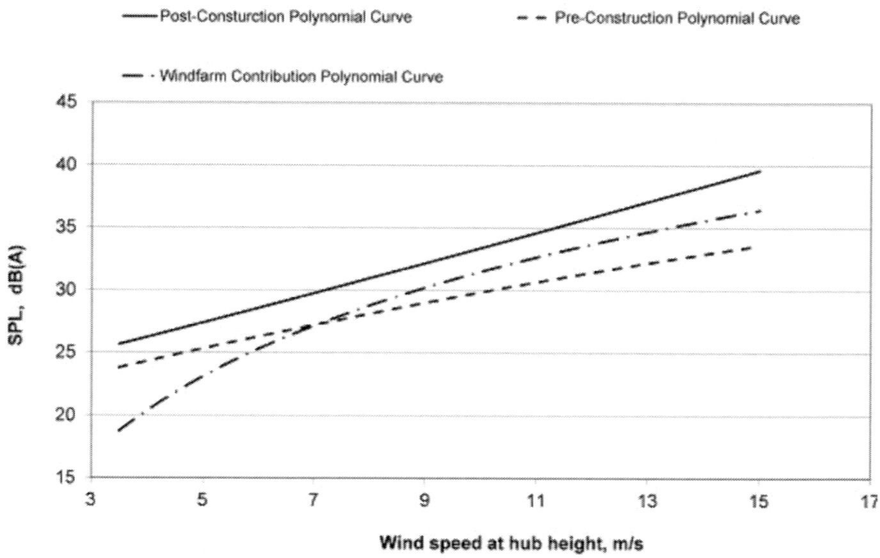

Figure 3. Implementation of correction for background procedure at a receiver next to a wind farm (L_{A90} descriptor, downwind).

3.3.3. Start and Stop Method

There may be doubts about the validity of the pre-construction background data, for example in the cases when total noise is below the background levels at particular wind speeds. Consequently, the conventional correction for background procedure cannot be performed. Start/Stop routines may be utilized instead for these situations. The Start/Stop method requires the wind farm to be shut down for a certain amount of time during a long term monitoring program or attended noise measurements. This allows for a comparison of noise levels during shutdown with operational periods in similar weather conditions. This method also allows for a comparison with operating periods adjacent to the shutdown periods. This method is very effective to determine noise contribution of the wind farm if pre-construction background levels have changed significantly.

It is not always required to organize the full shutdown of a wind farm. If a noise sensitive receiver is affected by a group of turbines, it may be sufficient to only control the operation of that specific group of turbines [7]. Multiple wind farm shutdowns should be done for different weather conditions and wind directions to determine the contribution of the wind farm for multiple localities. Although this method may be effective, it could be a very expensive exercise as it requires the wind farm to be out of electricity generating mode for a prolonged period of time. All these shutdown periods may add up to being a very expensive routine for wind farm owners.

Another major issue with the start/stop method is that it is difficult to collect a sufficient number of data points for times during stop periods. Collecting 2000 valid points of 10 minute data is equivalent to approximately 2 weeks' worth of data. This means that the wind farm in question has to be shut down for a period of 2 weeks or more, which is not practical. Therefore, the results obtained by this method normally involve processing less data compared to previous methods.

Data post- process for the start/stop method is similar to that described in the previous section.

4. Other Advanced Noise Monitoring and Data Processing Techniques

Problems pertained to noise impact from wind farms have attracted public and media attention, resulting in a number of research projects that employed different methods to address the challenges of wind farm noise monitoring. Non- standard methods of noise assessment can be potentially utilized in situations where conventional noise assessment procedures do not bring conclusive results.

4.1. Modified Total Noise and Background Correction Method

In the case where the total measured noise from a wind farm exceeds applicable noise criteria, conventional correction for background procedure is the standard tool to extract the wind farm contribution using a statistical approach. Actual background corresponding to the period when the total noise was acquired may be notably different from the pre- construction background due to season variations in the background, presence or absence of other ambient noise sources in the area, change of the wind speed reference or other factors. Consequently, magnitudes of the total noise and background noise may be very close at particular wind speeds or sometimes background levels can exceed the total noise. In these cases, the correction for background procedure cannot give a valid result or the accuracy of the estimates is doubtful.

One of the ways to address this issue is to update the background during the total noise data measurements or shortly before or after the data acquisition period. Frequently, such updates are not feasible from practicability and economic rationales. The method suggested in [9] can be utilized as a relatively simple procedure to resolve the situation from a statistical perspective. The method can be implemented in the steps summarized below.

Data for total noise measured at the receiver and background are considered within their wind speed bins. When the number of data points in the bins is too high, it is convenient to separate them into class intervals. 12-20 class intervals per wind speed bin is sufficient for practical tasks, where lower numbers of the class intervals corresponds to few tens of data in the bin and the higher number of the intervals is recommend for few hundreds of the data points. As a result of data segregation for a wind speed bin, we have the *i-th* class interval corresponding to measured total noise level L_i and the *j-th* class interval for the same wind speed bin corresponding to background level L_j. The probability of the event for a combination of levels L_i and L_j is:

$$p_{ij} = p_i p_j,$$

where p_i is the probability corresponding to the *i-th* total noise level and p_j is the probability corresponding to the *j-th* background level in the wind speed bin. In practical applications the probabilities are replaced by the frequencies of occurrence. The probabilities of all possible combinations for the total and background noise levels will form matrix *M*×*N* where *M* and *N* are numbers of interval classes for total and background noise respectively:

$$[p_{ij}]_{M\times N} = [p_i]_{M,1} \times [p_j]_{1,N},$$

where the string matrix $[p_i]$ is multiplied by row matrix $[p_j]$. The sum of the elements of the probability matrix must meet condition:

$$\sum_{i,j=1}^{M,N} p_{ij} = 1\ .$$

Events when the background noise exceeds the total noise are considered improbable and probabilities p_{ij} for such events must be set to zero. This delivers us a rectified probability matrix $[p_{i,j}^*]$ where the condition above is not met since the probabilities for events where the background is above the total noise are replaced by zeroes. Assuming that the rectified array of data represents all possible combinations of total noise and background pairs, the rectified probability matrix must be scaled by factor α:

$$\alpha = 1/\sum_{i,j=1}^{M,N} p_{ij}^*\ .$$

Wind farm contribution for events excluding the improbable is calculated following conventional rules for energy subtraction as explained in Section 3.1:

$$L_{i,j}^* = 10\ Log(10^{\frac{Li}{10}} - 10^{\frac{Lj}{10}}).$$

Wind farm contribution for the wind speed can be calculated as a mathematical expectation of the rectified set of data:

$$L_{WF} = \alpha \sum_{i,j=1}^{M,N} L_{i,j}^{*} p_{ij}^{*}.$$

If necessary, the standard deviation can be calculated for the wind speed bin:

$$s = \sqrt{\alpha \sum_{i,j=1}^{M,N} (L_{WF} - L_{i,j}^{*})^2 p_{ij}^{*}} .$$

It should be noted that in general, the calculated values of the wind farm noise L_{WF} in the wind speed bin can only rarely be approximated by the Gaussian probability distribution. Therefore, it is difficult to derive a universal relation for 95% confidence limits for the wind farm noise estimate. The standard deviation (or variance) can still be used as a measure of the data scattering.

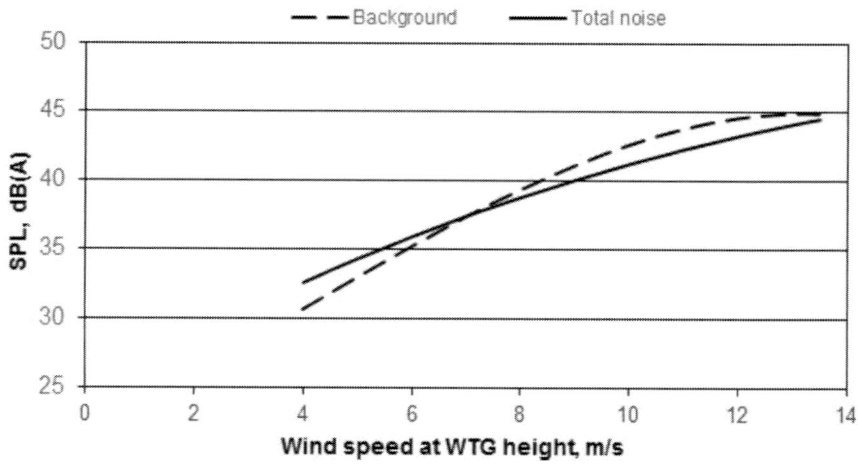

Figure 4. Total noise and background curves at a wind farm locality (L_{A90} descriptor, downwind).

Figure 4 shows a case where the standard correction for background method cannot be applicable to derive the wind farm noise for the full range of wind speeds. The background curve derived in accordance with the method described in Section 3.2 exceeds magnitudes of the total noise obtained using the similar procedure. Post- processing of the data in the wind speed bins does not improve the situation. The method based on the elimination of improbable combinations of total and background noise gives valid estimates in this case and can be utilized where updates of the background levels are not deemed feasible.

Table 1. Estimates of wind farm noise based on different data post-processing techniques (L_{A90} descriptor)

Wind speed, m/s	**4**	**5**	**6**	**7**	**8**	**9**	**10**	**11**	**12**	**13**
Curve fitting, dB(A)	28.2	28.7	27.9	20.6	N/A	N/A	N/A	N/A	N/A	N/A
Wind speed bins, dB(A)	27.4	30.4	28.2	N/A	N/A	N/A	N/A	N/A	N/A	44.3
Probable events, dB(A)	30.5	32.4	34.2	35.8	36	35.5	38.2	37.2	39.1	38.3

4.2. Start/Stop Method without Conclusive Difference in Levels

Sometimes even an organized Start/Stop event may not bring conclusive results to enable reliable calculations of wind farm noise contribution due to insignificant difference in the noise levels. Statistical analysis tools can be engaged to work out whether a wind farm noise may be derived based on data acquired during periods of wind farm operation and idling. The statistical test for equivalency of the means can be utilized to identify whether the correction for background methods or the method suggested in Section 4.1 can be used for wind farm noise calculations. This procedure was described in work [28] and was based on the analysis of an array of acoustical descriptors in the wind speed bins.

The procedure involves computations of the energy average in the wind speed bin as described in Section 3. The energy averages for the operating and shutdown periods are compared and if the difference does not exceed 2-3 dB, it is recommended to perform a check of the hypothesis about the equivalency of the means. If the null hypothesis is proven to be valid, i.e., the means can be considered equivalent, the acquired data does not provide a good basis for the wind farm noise calculation. A test based on Student's *t*-distribution (with equivalent or non-equivalent variance) with a confidence level 0.05 is a good tool to decide on the mean difference. A statistical test for equivalency of the variances (or standard deviations) may facilitate decision whether the operational and shutdown data are different enough from a statistical perspective. *F*-test for the standard deviations comparison is a widely used technique for comparison of statistical parameters of two data sets.

Table 2 shows a comparison of noise at a sensitive receiver acquired during operational and shutdown periods of a wind farm located in the adjacent area. Data compared are collected at similar wind speeds and directions. The mean values of the A- weighted SPLs do not differ significantly for wind speeds 8-10 m/s, as the difference is less than a half of decibel. However, the results of the test for equivalency of the means and standard deviation are dominantly negative. This is not surprising for the wind speeds at 6-7 m/s since the noise level difference is substantial at these wind speeds. However, the results of these tests indicate that operation of the wind farm is the reason for the statistical difference between the two arrays of data acquired during operational and shutdown periods. Consequently, the contribution of the wind farm may be calculated using the conventional energy subtraction method in the case of a significant difference in the levels. If the difference is subtle, it is better to utilize the computation routine described in Section 4.1. Only data at 9 m/s may not bring conclusive results since the test for the equivalency of the means is positive and may require additional data collection to bring a conclusive result.

Table 2. Comparison of noise levels acquired at wind farm shutdowns and similar operating periods (L_{A90} descriptor) and the results of relevant statistical tests

Wind speed, m/s	**6**	**7**	**8**	**9**	**10**
Shutdown SPL, dB(A)	23.3	28.7	35.1	37.1	38.6
Operational SPL, dB(A)	33.6	33.6	35.4	37.4	38.3
Equivalency of the means	Negative	Negative	Negative	Positive	Negative
Equivalency of the variances	Negative	Negative	Negative	Negative	Negative

4.3. Directional Noise Monitoring

Advancements in noise monitoring techniques enable the development of alternative methods for the monitoring of wind farms and other environmental noise sources. Advanced post-processing of signals from an array of microphones allows for the calculation of noise contribution from a particular angular direction. Currently, there are commercially available instruments that are able to do simultaneous data acquisition from three or more microphones [29, 30].

The directional contribution may be representative of wind farm noise when wind turbine locations can be localized within certain angles. In simple cases, as it is shown in Figure 5, a wind farm contribution can be considered to be equivalent to the directional contribution from the single sector:

$$L_{WF} = L(\alpha_1 < \alpha < \alpha_2) \, .$$

If there are several potential angles from which the wind farm noise can propagate, total contribution from the multiple angle directions can be identified by using rules of the energy summation. In some cases, noise contribution from a group of turbines deployed at significantly greater separation distance can be disregarded when there are turbines that are much closer. Consequently, the angle range can be narrowed to concentrate on a particular direction and provide better accuracy for the wind farm noise detection. There should also be no other significant noise sources in the angle sectors designated for the wind turbines.

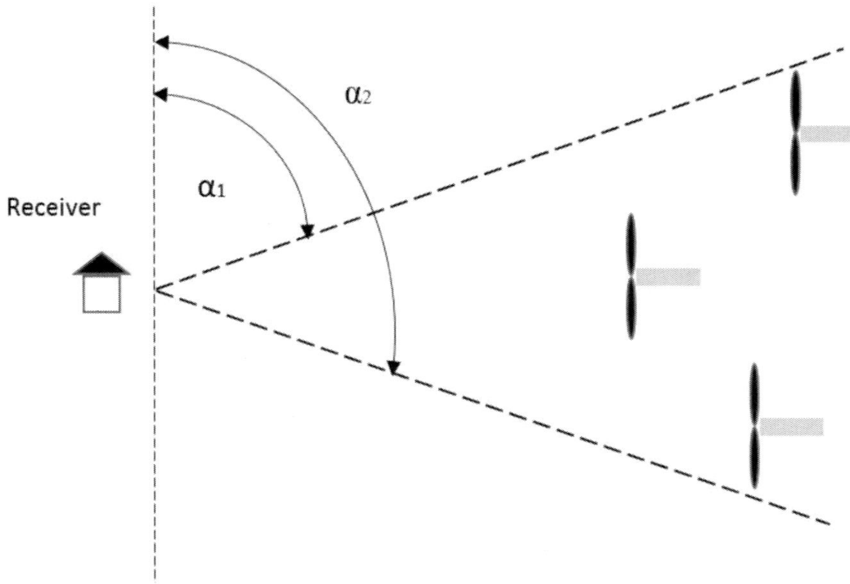

Figure 5. Scheme for directional noise monitoring.

Figure 6 shows an estimate of a wind farm noise based on directional levels corresponding to locations of the wind turbines [31] versus wind speeds between the cut-in wind speed and speed of the rated power. The directional contribution was acquired using a Barn Owl directional noise monitor [29]. The figure also shows the polynomial fit for A-weighted estimates of the signals low pass filtered with 1000 Hz cut-off frequency. Both of the fits are close to each other, which highlights the importance of mid-low frequency spans in wind farm noise evaluation for receivers located at a significant separation distance. The high frequency noise is attenuated and does not have a significant contribution to the wind farm noise magnitudes. Although the low pass filtered fit may cause a minor underestimate of the wind farm SPL, the filtered signals can be utilized to effectively eliminate interference from extraneous noise such as insect buzzing, which is typically high frequency in nature.

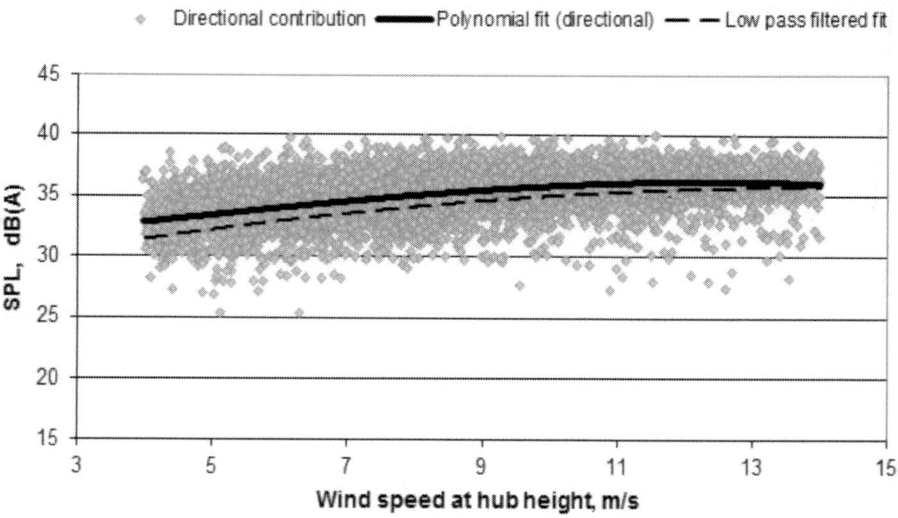

Figure 6. Wind farm noise as a directional contribution with polynomial fits for L_{Aeq} and low pass filtered signal (1000 Hz cut-off frequency).

4.4. Notes on Alternative Techniques

Continuous developments in areas of acoustic instrumentation and signal processing provide tools for more accurate and reliable detection of wind turbine noise in a complex acoustic environment. Time delay signal processing or beam forming techniques are now available in commercial instruments for directional noise monitoring and remote sensing. Three dimensional intensity measurements can be potentially utilized for tasks such as the measurement of sound power of wind turbines. It may also help to overcome disadvantages of a method where the sound power of a turbine is reported on the basis of SPL measurements at a single point [3].

Broader implementation of non- conventional methods of wind farm noise monitoring has significant predicaments such as the absence of relevant standards and insufficient information on the accuracy of the methods. This does not include the high cost of some advanced monitoring instruments and the relevant software required. These methods can be utilized for research and development projects. However, the implementation of such non- conventional monitoring methods for compliance checking purposes may be limited or even unacceptable. The situation may change in the future with the development of new instruments and noise monitoring methods. This growth could be further improved if additional scientific and regulatory support is provided.

GLOSSARY

A-weighting	Frequency weighting as specified in *Standard IEC 61672-1 Electroacoustics - Sound level meters - Part 1: Specifications* or its equivalent
Amplitude Modulation	A sound which is clearly perceived as rising and falling. It has varying, fluctuating noise character that is clearly audible above everything else.
Audio frequencies	Noise most audible to the general human ear, normally lies within the frequency range between 20 Hz to 20 kHz.
C-weighting	Frequency weighting as specified in *Standard IEC 61672-1 Electroacoustics - Sound level meters - Part 1: Specifications* or its equivalent
Decibel (dB, dB(Z) or dB(Lin))	Unit of sound pressure level, referenced to 20 μPa. Where dB is used in this chapter, it refers to the level in decibels with no frequency weighting applied.
G-weighting	Frequency weighting as defined in ISO 7196:1995, used to approximate how the human ear responds to infrasonic noise levels.
Equivalent noise level	The equivalent continuous A-weighted sound pressure level over the measurement time interval. Expressed as $L_{Aeq,\,T}$, where T refers to the measurement time interval.
Frequency	Rate of sound pressure variations – noise or sound is

	composed of energy across a wide range of frequencies including 20Hz or lower (infrasound).
Infrasound	Sound or noise where the energy lies mainly in the frequency range below 20Hz.
L_A, dB(A)	Unit of sound pressure levels which have had the A-weighting applied to them
L_C, dB(C)	Unit of sound pressure levels which have had the C-weighting applied to them
L_{den}, dB(A)	A-weighted SPL calculated for 24 hrs period with a 5 dB(A) penalty added to evening noise and a 10 dB(A) penalty added to night time measurements.
L_G, dB(G)	Unit of sound pressure levels which have had the G-weighting applied to them.
$L_{pA,LF}$, $L_{Aeq(LF)}$	Low frequency noise descriptor based on the A-weighted noise level calculated for the 1/3 octave band levels from 10Hz to 160Hz inclusive.
L_{A90}	A-Weighted Noise level which was equalled or exceeded for 90% of the measurement period. Typically used to represent the background noise level in an environment.
L_{Aeq}	Equivalent noise level – energy averaged A- Weighted noise level over the measurement period. The most common descriptor used to quantify environmental noise sources.
$L_{Aeq,10min}$	Equivalent A-Weighted noise level over a 10 minute measurement period.
Noise Floor	The minimum noise level which can be recorded by a specific measurement instrument.
Sound Pressure Level (SPL)	A logarithmic measure of effective sound pressure relative to a reference value (normally 20 μPa). Measured in dB.
Tonal noise	Noise with a perceptible and definite pitch or tone.

REFERENCES

[1] European Environment Agency, *Good practice guide on noise exposure and potential health effects, EEA, Copenhagen*, 2010.

[2] World Health Organization, *Night noise guidelines for Europe*, WHO, Bonn, 2007.

[3] International Electrotechnical Commission IEC 61400-11 Edition 3.0, Wind turbines- Part 11: Acoustic noise measurements techniques, 2012.

[4] Haugen, KMB. International Review of Policies and Recommendations for Wind Turbine Setbacks from Residences, Minnesota Department of Commerce, 2011

[5] Arrêté du 26 août 2011 relatif aux installations de production d'électricité utilisant l'énergie mécanique du vent au sein d'une installation soumise à autorisation au titre de la rubrique 2980 de la législation des installations classées pour la protection de l'environnement, NOR: DEVP1119348A, La ministre de l'écologie, du développement durable, des transports et du logement.

[6] Koppen, E; Fowler, K. International legislation for wind turbine noise, *Proceedings of Euronoise*, 2015, Maastricht, May-June 2015.

[7] Standards New Zealand NZS 6808, *Acoustics- Wind farm noise*, 2010.

[8] SA EPA, *Wind farm environmental noise guidelines*, Government of South Australia, Adelaide, 2009.

[9] Lenchine, VV. Statistical and curve fitting post- process of wind farm noise monitoring data, *Proceedings of International Congress on Acoustics*, Sydney, August 2010.

[10] Leventhall, G. *A Review of Published Researched on Low Frequency Noise and its Effects*, report prepared for DEFRA, London 2003.

[11] Department of Environment, Food and Rural Affairs, *Proposed criteria for the assessment of low frequency noise disturbance*, report prepared by University of Salford, DEFRA, London 2005.

[12] German Institute for Standardization DIN 45680, *Measurement and evaluation of low-frequency environmental noise*, 1997.

[13] Broner, N. A simple outdoor criterion for assessment of low frequency noise emission assessment, *Journal of low frequency control, vibration and active control*, 29 (1), (2010), pp 1–13.

[14] *Statutory Order on Noise from Wind Turbines*, The Danish Ministry of Environment, Copenhagen, revision 2011.

[15] Poulsen, T; Mortensen, F. *Laboratory Evaluation of Annoyance of Low Frequency Noise*, Danish Environmental Protection Agency, Copenhagen, 2002.

[16] Tachibana, H; Yano, H; Fukushima, A; Sueoka, S. Nationwide field measurements of wind turbine noise in Japan, *Journal of Noise Control Engineering*, 62(2), (2014), pp. 90-101.

[17] Wind turbine amplitude modulation: research to improve understanding as to its cause and effect, Renewables UK, London, UK, 2013.

[18] Lenchine, VV. Variations in sound pressure levels under random change of atmospheric conditions, *Proceedings of Acoustics, 2012*, Freemantle, November 2012.

[19] Lenchine, VV; Song, J. Special noise character in noise from wind farms, *Proceedings of International Congress on Noise Control Engineering- Internoise, Melbourne*, November 2014.

[20] International Organization for Standardization ISO 7196, Acoustics- Frequency- weightings characteristic for infrasound measurements, 1995.

[21] Jakobsen, J. Danish guidelines on environmental low frequency noise, infrasound and vibration, *Journal of Low Frequency Noise, Vibration and Active Control*, 20 (3), (2001), pp. 141–148.

[22] Yamada, S; Ikuji, M; Fujikata, S; Watanabe, T; Kosaka, T. Body sensations of low frequency noise of ordinary persons and profoundly deaf persons", *Journal of Sound and Vibration*, 2(1), 1983, pp. 32–36.

[23] Victoria Department of Health, *Wind farms, sound and health*, Government of Victoria, Melbourne, 2013.

[24] National Medical and Health Research Council, Evidence on wind farms and human health, NHMRC, Canberra, 2015.

[25] Evans, T; Cooper, J; Lenchine, V. *Infrasound levels near wind farms and in other environments*, EPA, Adelaide, 2013.

[26] Wang, L; Zander, AC; Lenchine, VV. Measurement of the self-noise of the microphone wind shields, *Proceedings of Australasian Fluid Mechanics Conference*, Launceston, December 2012.

[27] International Electrotechnical Commission IEC 61672.1 Ed.2.0, Electroacoustics- Sound level meters, Part 1: Specifications, 2013.

[28] Lenchine, VV; Song, J. Infrasound and blade pass frequency levels in areas adjacent to wind farms, *Proceedings of International Congress on Noise Control Engineering- Internoise*, Melbourne, November 2014.

[29] Bullen, R. Long-term environmental monitoring and noise source identification, *Acoustics Australia*, 31(1), 2003, pp.23-27.

[30] Acoustic Research Laboratories, Environmental noise compass (brochure), ARL, Australia, accessed 21st October 2015, http:// www.acousticresearch.com.au/files/ENC_brochure_v2.pdf.

[31] Lenchine, VV; Holmes, B. Contribution of a wind farm noise into noise at a distant receiver in a rural environment, *Proceedings of Acoustics*, 2011, Gold Coast, November 2011.

BIOGRAPHICAL SKETCH

Name: Valeri Lenchine

Affiliation: Science, Assessment & Planning Division, Environment Protection Authority, South Australia

Education: Ph.D.- Acoustics & Vibration (Mechanical/Aerospace Engineering), Samara State Aerospace University; BS & MS (high distinction), Mechanical/Aerospace Engineering, Samara State Aerospace University

Address: office- EPA, GPO Box 2607 Adelaide SA 5001 Australia

Research and Professional Experience:

Work history:

- Current: Principal Adviser and Noise & Vibration team leader, Science, Assessment & Planning Division, Environment Protection Authority, Australia (since 2008). Project and team management, consulting, development of regulatory documents and procedures, performing test and monitoring programs, R&D projects.
- Previous industry appointments: senior engineering positions in VIPAC Engineers and Scientists (Australia, 2006~ 2008) and Samsung Electronics (South Korea, 2002-2005). Duties included project management and execution, consulting, laboratory and field tests, product development and designing.
- Previous academic appointments: Associate Professor at Samara State Aerospace University (last appointment, 1991-2002), Visiting Fellow at Yale University (1999-2000), Head of Acoustic Measurement Lab at Machine Acoustics Institute (last appointment, 1995-2002).

Main projects in industry, last 10 years:

- Wind farms and renewable energy plants.
- Assessment of major energy, mining and infrastructure projects (Ceres Wind Farm, Olympic Dam expansion etc).
- Prediction and assessment of vibration and noise from rail operations and transport corridors.
- Rotor dynamics and reliability analysis for turbo- generator units and supporting machinery in power generation, vibration/noise abatement solutions in industry.

- Development of test facilities and programmes for aero-space, power generation, automotive and defense applications.
- Performance, reliability, noise and vibration problems in commercial air-conditioning systems and its elements.

Coauthor and author of 16 patents and patent applications registered in the USA, South Korea, Japan and China.

Professional Appointments:

- Member of the Acoustical Society of America, Australian Acoustical Society, Vice- chairman of South Australian Technical Committee.
- Recognition as Professional Engineer by the Institution of Engineers Australia
- Registered reviewer for acoustic journals and relevant editions.
- Member of former Australian Environment Protection & Heritage Council working group, other advisory boards and committees.
- Member of organizing committees for International and national acoustic conferences.

Publications Last 3 Years:

1. Lenchine V.V., Assessment of amplitude modulation in environmental noise measurements//Applied Acoustics, Vol.104, March 2016.-P.152-157.
2. Alamshah S.V., Zander A.C., Lenchine V.V., Effects of turbulent flow characteristics on wind- induced noise generation in shielded microphones//Proceedings of AAS conference Acoustics-2015 Hunter Valley, Australia, 2015.
3. Lenchine V.V., Song J., Special noise character in noise from wind farms//Proceedings of Internat. Conference Internoise-2014 Melbourne, Australia, 2014.
4. Lenchine V.V., Song J., Infrasound and blade pass frequency levels in areas adjacent to wind farms//Proceedings of Internat. Conference Internoise-2014 Melbourne, Australia, 2014.
5. Lenchine V.V., Song J., Assessment of rail noise based on generic shape of the pass-by time history//Proceedings of AAS conference Acoustics-2013 Victor Harbor, Australia, 2013.

6. Alamshah S.V., Zander A.C., Lenchine V.V., Development of a technique to minimise the wind- induced noise in shielded microphones//Proceedings of AAS conference Acoustics-2013 Victor Harbor, Australia, 2013.
7. Evans T., Cooper J., Lenchine V., Infrasound levels near wind farms and in other environments, EPA, Adelaide, 2013.
8. Lenchine V.V., Variations in sound pressure level under random change of atmospheric conditions//Proceedings of AAS conference Acoustics-2012 Fremantle, Australia, 2012.
9. Wang L., Zander A.C., Lenchine V.V., Measurement of self- noise of microphone wind shields//Proceedings of 18th Australasian Fluid Mechanics Conference, Launceston, Australia, 2012.

Name: Jonathan Song
Affiliation: Science, Assessment & Planning Division,
Environment Protection Authority, South Australia
Date of Birth: 26-12-89
Education: University of Adelaide, B.Eng (Mechatronics)
Address: office- EPA, GPO Box 2607 Adelaide SA 5001 Australia

Research and Professional Experience:
Work history:

Science, Assessment & Planning Division, Environment Protection Authority, Australia, March 2012~Present. Performing tests and monitoring programs, R&D projects, data analysis.

Main projects in industry:

- Assessment of environmental impact from new and existing wind farm developments
- Infrasound noise monitoring study
- Data processing and analysis of long term noise monitoring projects
- Assessment and prediction of noise from industries, infrastructure and transport corridoors

Publications Last 3 Years:

1. Lenchine V.V., Song J., Special noise character in noise from wind farms //Proceedings of Internat. Conference Internoise-2014 Melbourne, Australia, 2014.
2. Lenchine V.V., Song J., Infrasound and blade pass frequency levels in areas adjacent to wind farms //Proceedings of Internat. Conference Internoise-2014 Melbourne, Australia, 2014.
3. Lenchine V.V., Song J., Assessment of rail noise based on generic shape of the pass-by time history //Proceedings of AAS conference Acoustics-2013 Victor Harbor, Australia, 2013.

In: Wind Farms
Editor: Marian Dunn
ISBN: 978-1-63484-841-1

Chapter 3

POWER QUALITY OF OFFSHORE WIND FARMS: MEASUREMENT, ANALYSIS AND IMPROVEMENT

Qiang Yang*
College of Electrical Engineering, Zhejiang University,
Hangzhou, China

ABSTRACT

In recent years, with the quick development of offshore wind farms, there is an urgent and increasing demand on investigating the power quality of grid-connected offshore wind farm and understanding its impacts on the operation of power grid. This chapter focuses on addressing the aforementioned technical challenges and exploits the power quality issues of offshore wind farms from a number of aspects to enable us to model, analyze and protect the power quality of large-scale offshore wind farms. This chapter explores the modeling approach of semi-aggregated equivalent model of offshore wind farm based on PSCAD/EMTDC, which can be adopted for the study of measurement, analysis and improvement of power quality at point of common connection (PCC). Following to this, this chapter attempts to address this technical challenge through a simulation-based study by the use of PSCAD/EMTDC models and carries out an assessment of power quality at the Point of Common Coupling (PCC) in the scenario of offshore wind

* E-mail: qyang.

farm integrated into the power network whilst reduce the impact of index discrepancy and uncertainty. Finally, considering the integration of hybrid energy storage system (HESS) including battery energy storage system (BESS) and super-capacitors energy storage system (SCESS) to improve the power stabilization in power grid, the control strategy on managing the HESS to stabilize the power fluctuation in a real-time fashion without the need of predicting wind speed statistics is also presented. The suggested solutions are assessed through a set of simulation experiments and the result demonstrates the effectiveness in the simulated offshore wind farm scenarios.

Keywords: offshore wind farms, power quality, HESS, BESS, wind farm model, HVAC, SVC

INTRODUCTION

It is envisioned that the total global installed generation capacity will be increased from 319.6GW in 2013 to 678.5GW in 2020. In general, as the offshore wind can provide more consistent and strong wind, the offshore wind turbines can operate at higher capacity factors and received particular attentions in the past decade, and the offshore wind energy is expected to reach nearly 40GW in 2020.

Due to the urgent demand on utilizing renewable power generation resources and low-carbon power supply across the overall world, much research and engineering effort has been made to investigate the potentials of renewable energy, e.g., wind and solar energy, and the wind energy sector has become one of the most important players in the renewable energy markets. However, due to the intermittent nature of the wind generators and the tower-shadow effect imposed on the downwind turbines in the offshore wind farm, the electrical power delivered by this type of generation possesses similar intermittent and fluctuant features. Furthermore, wind turbines are usually integrated with the grid at remote terminals, far from central loads or conventional generation, as shown in Figure 1.

This chapter focuses on addressing the aforementioned technical challenges and exploits the power quality issues of offshore wind farms from a number of aspects to enable us to model, analyze and protect the power quality of large-scale offshore wind farms. The chapter is organized as follows:

Firstly the modeling approach (semi-aggregated equivalent model) of offshore wind farm based on PSCAD/EMTDC is presented.

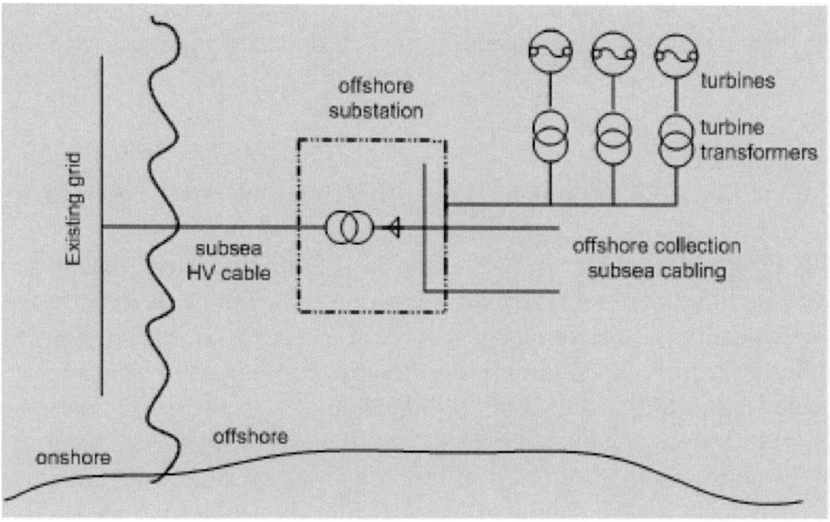

Figure 1. The illustration of the offshore wind farms.

As the size and number of wind power farms penetrated in power system increase, the accurate model of the wind farm is more important for the research on the influence of large-scale wind farm on power system. The complete model of wind farm with a large number of wind turbines, including the modeling of all the wind turbines and the internal electrical network, presents high-order and costs long computation time when simulation. In order to reduce the model order and computation time, equivalent models have been developed to represent the collective response of the wind farm at PCC to grid;

In addition, the measurement and analysis of power quality for grid-connected offshore wind farm is one of the key issues. This chapter attempts to address this technical challenge through a simulation-based study by the use of PSCAD/EMTDC models. This chapter carries out an assessment of power quality at the Point of Common Coupling (PCC) in the scenario of offshore wind farm integrated into the power network whilst reduce the impact of index discrepancy and uncertainty.

Finally, the integration of hybrid energy storage system (HESS) including battery energy storage system (BESS) and super-capacitors energy storage system (SCESS) has been considered one of the appropriate solutions to meet the technical challenge of power stabilization in power grid consisting of renewable generations. The control strategy on managing the HESS to stabilize the power fluctuation in a real-time fashion without the need of predicting wind speed statistics is also presented. The suggested solutions and control strategy are assessed through a set of simulation experiments and the result demonstrates the effectiveness in the simulated offshore wind farm scenarios.

EQUIVALENT MODEL OF OFFSHORE WIND FARMS

This section exploits the establishment of an equivalent model for a 200MW offshore wind farm under construction and research the improvement of power quality at point of common connection (PCC) on PSCAD/EMTDC. Considering that the wind turbines in the different regions of wind farm are clustered into groups according to the received wind speed, the semi-aggregated equivalent model, not the aggregation into a single wind generator, is applied to represent the collective response of the offshore wind farm based on the paralleled transformation of internal collector network. Compared with constant speed constant frequency (CSCF) wind generator, variable speed constant frequency (VSCF) doubly fed induction generator (DFIG) presents noticeable advantages such as, the optimum tip-speed ratio to get maximum wind-power, the achievement of decoupled control of active and reactive powers, and the less power demand of converter [1-3]. As a result, DFIG has become the most widely used wind turbine for wind farms. In the offshore wind farm under construction, there are 40 5MW DFIGs. We can believe that DFIG will have significant applications in the future development of Chinese offshore wind farms.

The complete model of wind farm with high number of wind turbines, including the modeling of all the wind turbines and the internal electrical network, presents high-order and costs long computation time in numerical simulations. In order to reduce the model order and computation time, equivalent models have been developed to represent the collective response of the wind farm at PCC to grid [4-6]. The wind turbine generators in the offshore wind farm are connected by the internal electrical network. In some existing research, the internal electrical network is either ignored or replaced by equivalent impedance [4, 5]. Considering that in the wind farm the wind turbines receiving the similar incoming wind vary, do not keep unchanged, the paralleled transformation method for the internal radial collector network of wind farm is applied in [7], which can make any two wind turbines in parallel and then make it possible to aggregate the wind turbines in any positions according to the wind situation. Based on the paralleled transformation of internal collector network of wind farm, the semi-aggregated equivalent method is presented to model the offshore wind farm, i.e., the wind turbines with the similar incoming wind are aggregated to a single wind turbine.

Semi-Aggregated Equivalent Modle

DFIG wind turbine includes a rotor induction generator connected to the wind turbine rotor through a gearbox. The stator of DFIG is directly connected to the grid whereas the rotor winding is connected with two back-to-back voltage source converters. By adjustment of the switching of the IGBT in both converters, the power flow between the rotor circuit and the grid can be controlled both in magnitude and in direction [8]. Rotor side converter usually provides active and reactive power of the machine while the grid side converter keeps the voltage of the DC circuit constant [9]. In general, the DFIG turbines can be organized in three different patterns (bus structure, radial structure and hybrid structure) in the wind farms, as shown in Figure 2.

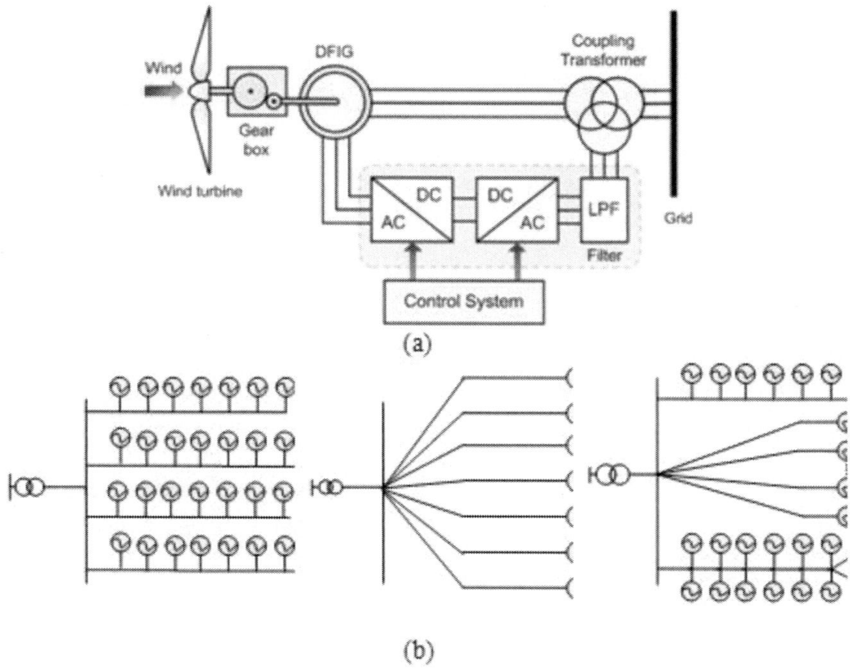

Figure 2. (a): The schematic diagram of DFIG wind turbine; (b) DFIG patterns in wind farms.

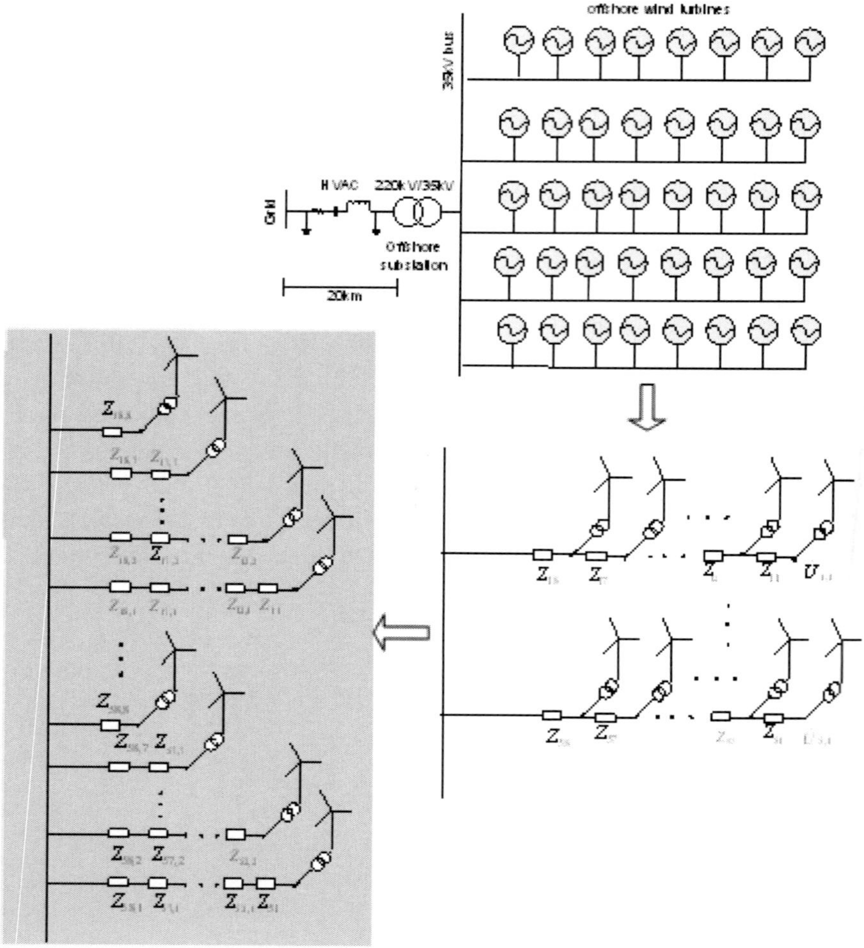

Figure 3. The layout of offshore wind farm and the parallel transformation.

In order to exploit the impact of a wind farm on the connected power grid, equivalent model has to been developed to represent the collective response of the wind farm at PCC to grid, instead of the complete model including the modeling of all the wind turbines [10].

Since the wind situation varies, it is common that the group of the wind turbines changes, which receive the similar wind speed. Thus it is necessary to establish the equivalent model of internal electrical network to make the wind turbine in parallel with each other, which can bring about an advantage that in aggregating the wind turbines with similar wind speed we do not need to care about the specific positions of those wind turbines. The layout of offshore wind farm and the parallel transformation is given in Figure 3.

For the internal collector network in the offshore wind farm, the method of parallel connection aimed transformation is applied to develop the equivalent model, which is based on the principle that the wind turbine terminal voltages produced by grid voltage should keep unchanged before and after the transformation [7].

As shown in Figure 3, for each collector network, the transformation starts from the end of it, and then move step by step to PCC. The equivalent transformation is illustrated as follows: (1) calculate the steady-state equivalent impedance Z_n of the wind turbine generator branch i; (2) the line impedance $Z_{li}(i \geq 2)$ is divided into i parts, denoted as $Z_{li,n}(n = 1,...,\mathrm{i})$, and then series the average impedance $Z_{li,n}$ to the original branch of the wind turbine generator; (3) amend the impedance of the wind turbine generator branch where the paralleled transformation has been finished, as $Z_n + Z_{li,n}$; Implement the same parallel transformation on the next line impedance $Z_{l(i+1)}$ until the entire radial collector network is done.

After the transformation, the wind turbine generators in the offshore wind farm will turn into completely parallel structure, which is well adapted to the aggregation of those with the similar wind speed. The offshore wind farm is modeled by one equivalent model with the semi-aggregated equivalent method, representing the entire offshore wind farm with highest possible accuracy. The offshore wind farm can be clustered into different groups based on the variable wind speed in different regions of the wind farm. The wind turbines experiencing the similar wind speed are arranged in a group and can be aggregated into an equivalent model with the weighted average method. As the group of wind turbine consists of wind turbines receiving similar incoming wind and generating the similar terminal voltage and rotor speed, the simple and effective equivalent method is the weighted average method [6]. Thus, the required equivalent p.u. values of the DFIG are stator resistance, stator reactance, rotor resistance, rotor reactance, excitation reactance and the generator inertia constant. They are calculated as the following:

$$X_{eq} = \frac{\sum_{j \in \forall G} S_j X_j}{\sum_{j \in \forall G} S_j}$$

where i represents the wind turbine; S is the wind turbine capacity; G denotes the wind turbine group and X represents the above mentioned p.u. parameters which need to be equivalently transformed.

The required equivalent actual values (i.e., not in per unit values), such as the rated capacity, the mechanical power and the electrical power of wind turbines, are equal to the sum of that of the individual wind turbines, calculated as follows:

$$S_{eq} = \sum\nolimits_{j \in \forall G} S_j$$

To verify the aforementioned semi-aggregated equivalent model, a set of simulation experiments are carried out based PSCAD/EMTDC (Power Systems Computer Aided Design) platform. The test scenario of 3 DFIG wind turbines (2MW) connected in a hybrid structure are used to evaluate and compare the performance between the equivalent model of an aggregated 6 MW wind turbine and the original 3-turbine system from a number of aspects, including active power, reactive power and transient performance under system failures.

Figure 4 and Figure 5 provide the simulation results for the steady-state output active power and reactive power performance for both original model and the equivalent model, respectively.

It can be seen that the proposed semi-aggregated equivalent model performs accurately in both steady-state active and reactive power output with the accepted error, implying that such approximated model can be used for analysis and evaluation of wind farms.

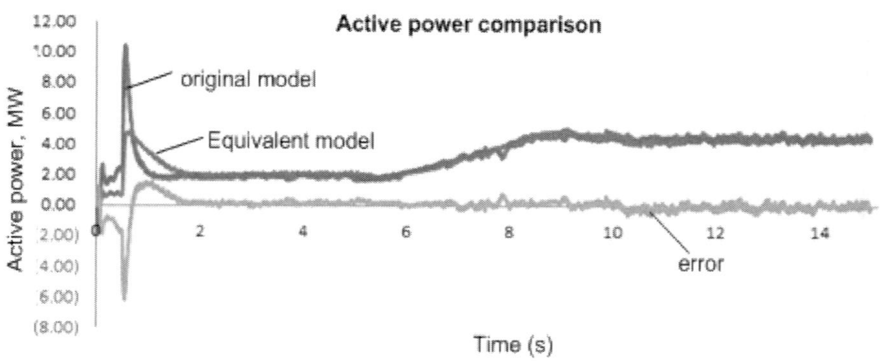

Figure 4. The comparisons of active power performance.

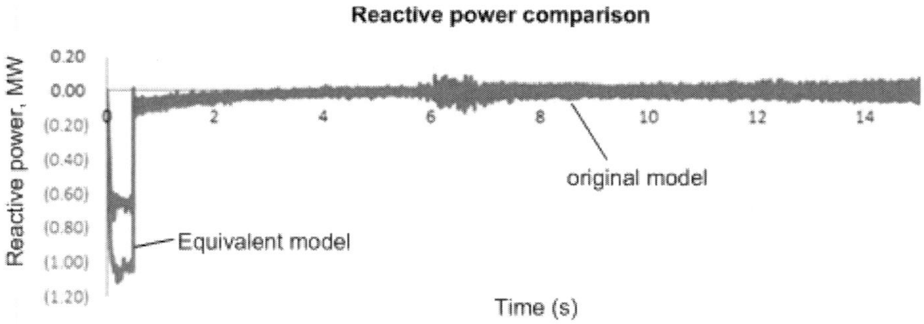

Figure 5. The comparisons of reactive power performance.

Now we look into the transient performance of the semi-aggregated equivalent model of the wind farms under the scenario of single-phase failures in comparison with the complete model.

Figure 6 presents the assessment of active power, reactive power and voltage for both two models. It clearly shows that, under the condition of single-phase failure, the semi-aggregated equivalent model can well match the transient characteristics of the complete model under the transient failure process.

Summary: In this chapter, the equivalent model of an offshore wind farm under construction and its transmission system is established on PSCAD/EMTDC, including the paralleled transformation based internal electrical network equivalent model and the semi-aggregated equivalent model of wind farm by aggregating the wind turbines with the similar wind speed into a single wind generator, HVAC transmission system and reactive power compensation SVC. The simulation results show that the equivalent model of offshore wind farm can represent the collective response of the wind farm effectively.

Power Quality Measurement and Analysis of Offshore Wind Farm

With the growth of environmental concerns, the VV massive increase in the prices of fossil fuels, and the rapid increase on the electric power demand, interest in renewable energy sources is increasing. Wind energy is one of the top growing renewable energy technologies in the world [11]. The offshore wind resource of our country is very abundant. Offshore winds blow stronger and more uniformly than on land, resulting in greater potential [12, 13].

However, due to the stochastic nature of the offshore wind, electrical power delivered by this type of generation possesses similar features. Furthermore, wind turbines are usually integrated with the grid at remote terminals, far from central loads or conventional generation [14]. Therefore, there is an urgent demand on investigating the power quality of grid-connected offshore wind farm and its impact on the operation of power grid. The output power of the wind turbine exhibits fluctuation and intermittent nature due to uncertainty of wind energy resources and the operation characteristics of the wind turbine, and hence it results in the problems like voltage fluctuation, voltage flicker, harmonics, voltage unbalance, voltage deviation, frequency deviation, etc. [15]. The power quality indices are used to quantify the quality of the generated and its supply to the grid. Resent years, much research effort has been made to study power quality disturbance caused by the grid-connected wind farms. In [16], it represented a real-time monitoring of power quality indexes of 100MW wind farm in Inner Mongolia. In [17], the study built the entire assessment system of wind farm in PSCAD/EMTDC and wind speed models based on the actual measured data. In [18], the authors calculated the power quality indices by investigating the island wind farm in Shantou power grid and made a synthetic evaluation of the power quality. In [19], the work examined the effect of the integration of wind energy to 110 kV power transmission network and analyzed the impact on the power quality on medium voltage (MV) and low voltage (LV) power networks.

In this section, an equivalent model with 50MW capacity for offshore wind farm is developed on PSCAD/EMTD. Static Var Compensator (SVC) is applied to compensate the reactive power generated. We focus on the measurement method of power quality and analysis of the effect of offshore wind farm when connected to the power grid. Simulation experiments are carried out to measure and calculate the indices of power quality at the point of common coupling (PCC) of the wind farms and the power system.

We consider the wind farm with the total capacity of 50MW. Figure 7 shows the system configuration of the wind farm to be analyzed. An equivalent model of wind turbines and internal electrical networks is obtained with the weighted average method. The interconnection to the onshore grid is ensured by submarine cables. The short-distance submarine cables are modeled as π lines. Power quality is calculated at speed of 11m/s. The rated power of DFIG is 5MW, rated voltage is 0.69kV. For the 50MW equivalent model, equivalent parameters are listed as stator resistance γ_s =0.0238pu, rotor resistance γ_r = 0.0194pu, stator self-inductance L_s = 0.294pu, rotor self–inductance L_r = 0.515pu, mutual inductance between stator and rotor L_m = 14.841pu. SVC is employed to compensate the reactive power mainly generated by the AC submarine cables.

Power quality indices are used to quantify the degree of the power quality disturbance. In this study, we consider the following four indices which are defined as follows.

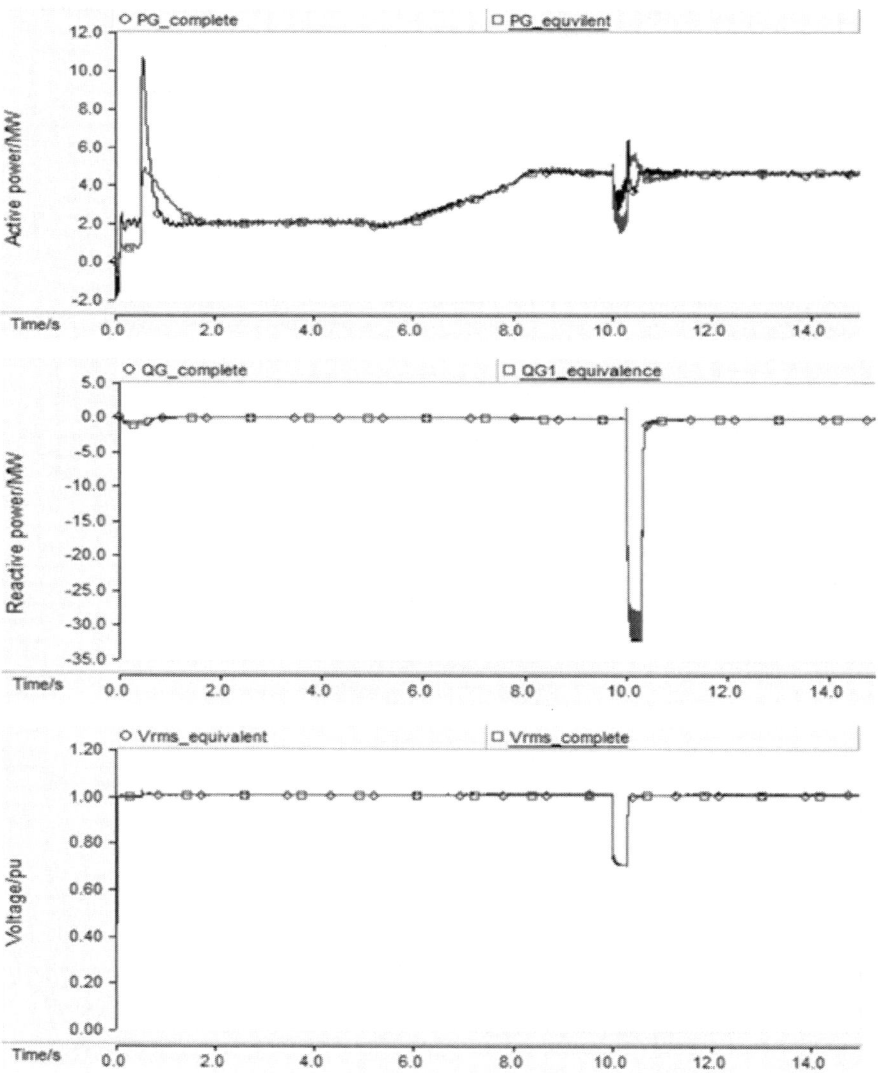

Figure 6. The comparisons of transient performance under single-phase faults.

Voltage Unbalance

Voltage unbalance is in general due to the large unbalanced power loads. The unbalance in voltage causes negative sequence currents to flow in Induction machines, causing overheating [11]. Three–phase unbalance degree is defined as the following:

$$\varepsilon = \frac{\sqrt{1-\sqrt{3-6L}}}{\sqrt{1+\sqrt{3-6L}}} \times 100\%$$

where $L = (a^4 + b^4 + c^4)/(a^2 + b^2 + c^2)$, a, b, c is the rms value of U_a, U_b, U_c.

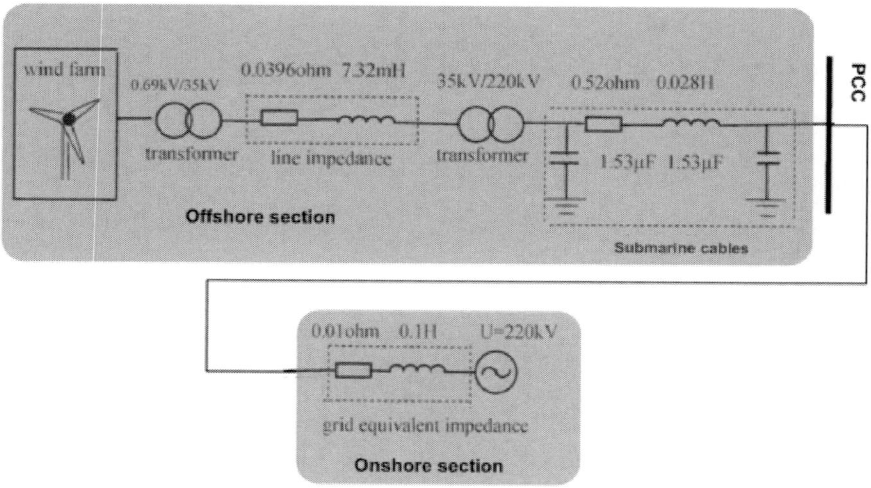

Figure 7. Grid-connected offshore wind farm system configuration.

Voltage Harmonics

Voltage harmonics is created by non-linear loads, such as transformers, rotating electric machines, and power electronics components. Voltage harmonics may result in over heating of transformers and generators. The voltage total harmonic distortion can be defined as follows:

$$THD_v = \sqrt{\sum_{h=2}^{N} \frac{V_h^2}{V_1^2}}$$

where V_h is the voltage *rms* value for the harmonic of order h.

Voltage Fluctuation and Flicker

Voltage fluctuation is caused by the fluctuation of output power due to wind speed variation. It may produce perceivable visual effects to human eyes through incandescent lamps, which can damage sensitive equipments [20]. Voltage fluctuation value d is calculated as follows:

$$d = \frac{\Delta U}{U_N} \times 100\% = \frac{U_{\max} - U_{\min}}{U_N} \times 100\%$$

where $U_{\max}$, U_{min} is the peak value of the half period *rms* value.

Flicker takes place when voltage fluctuates in a range of frequency between 0.05-35Hz. Short-term flicker severity P_{st} is adopted to evaluate the flicker severity measured over 10 minutes.

IEC recommended flicker model [21] includes five blocks: input voltage adaptor block, squaring multiplier block, weighting filters block which are composed of a first-order high-pass filter, a sixth-order Butterworth low-pass filter and the weighting filter which responds to the frequency content of the input waveform with an output that matches the human sensitivity response, squaring and smoothing block that simulates the non-linear eye-brain perception and the storage effect in the brain, on-line statistical analysis block.

National Limits of Power Quality Indices

Table 1 shows the national limiting values of the power quality indices at 220kV voltage level. This can be used as the reference to evaluate the power quality measurements obtained from our simulation experiments.

Table 1. Indexes of Power Quality and Limiting Value [22]

Power Quality Index	PARAMETER	Qualified Range
voltage fluctuation	$d/\%$	≤ 3
voltage flicker severity	P_{st}	≤ 0.8
harmonics	$THD/\%$	≤ 2
voltage unbalance	$\varepsilon/\%$	≤ 2

Flicker Measurement Based on Wavelet Transform

Voltage fluctuation and flicker are considered imposing major negative impacts on power quality [24]. The conventional voltage flicker measurement methods, e.g., the square demodulation, full wave commutation, and half wave commutation, are not appropriate to measure time-changing voltage flicker signal and analyze its time-frequency [24, 25]. In this study, the voltage signals at PCC are measured and flicker signal is obtained with the Mallet wavelet analysis.

S. Mallat presented the concept of multi-resolution analysis (MRA) when constructing orthogonal wavelet basis, given rapid algorithm of the constructional measure of orthogonal wavelet and orthogonal wavelet transformation, that is to say Mallat algorithm [26]. The basic idea behind the Mallat algorithm can be summarized as follows: Assuming that the approximation of $f(t) \subset L^2(R)$ in resolution 2^{-j} is C_j, then the approximation of $f(t)$ in resolution $2^{-(j+1)}$ named C_{j+1} can be got by filtering C_j with low pass filters derived from scale function $\varphi(t)$, and discrete detail D_j in the resolution between 2^{-j} and $2^{-(j+1)}$ can be got by filtering C_j with high pass filters derived from wavelet function $\psi(t)$ and scale function $\varphi(t)$. The decomposition of the signal in terms of approximation and detail version are given as:

$$c_{j+1}(k) = \sum_{n=-\infty}^{\infty} c_j(n) h_0(n-2k)$$

$$d_{j+1}(k) = \sum_{n=-\infty}^{\infty} c_j(n) h_1(n-2k)$$

where $h_0(k)$ has a low-pass filter response and $h_1(k)$ has a high-pass filter response. $c_j(k)$ is referred to the scale coefficient, $d_j(k)$ is referred to the wavelet coefficient.

The reconstruction algorithm is shown as below:

$$c_j(k) = \sum_{n=-\infty}^{\infty} h_0(k-2n)c_{j+1}(n) + \sum_{n=-\infty}^{\infty} h_1(k-2n)d_{j+1}(n)$$

Now we look in to the application of Mallet algorithm in flicker detection, as an example. The measured voltage is described as:

$$u(t) = A(t)\cos(\omega_0 t + \theta) = A\left[1 + \sum_k m_k \cos(\omega_k t)\right]\cos(\omega t + \theta)$$

Assuming that the modulating voltage is single,

$$u(t) = A\left[1 + m\cos(\Omega t)\right]\cos(\omega t)$$

Then,

$$u^2(t) \approx \frac{A^2}{2} + mA^2\cos(\Omega t) + \frac{A^2}{2}\cos(2\omega t)$$

where $m\cos(\Omega t)$ is the flicker signal. Mother wavelet, db4, is employed in this study. The sampling frequency is 4 kHz, in order to make the frequency range of approximation coefficient is between 0.05Hz and 35Hz, signal $u^2(t)/A^2$ is decomposed into 15 levels and reconstructed. The Mallet wavelet transform is used to replace the function of high pass filter and low pass filter in the IEC recommended flicker model. After we obtain the trend of the instantaneous flicker sensation, short-term flicker severity P_{st} can be calculated to with the discrete flicker meter model developed based on MATLAB.

Simulation Experiments and Numerical Results

We take the model presented in Section II. The wind speed is one of the key factors affecting the wind power. Wind speed is modeled on PSCAD by wind source module, including four components: mean speed, gust wind, ramp wind and noise wind. Based on the wind resources in the Lai Zhou bay offshore wind farm, 8m/s is chosen as the mean wind.

$$V_{wind} = V_m + V_g + V_r + V_n$$

Figure 8(a) shows the wind speed V_{wind}. The gust wind starts at 7s and lasts for 6s, with a peak gust wind speed 2m/s. The damp wind starts at 20s and lasts for 8s, with a peak damp wind speed 2m/s. The noise wind works the whole time. Figure 8 (b) shows the active power measured at PCC.

It can be seen that the system realizes the maximum wind power tracking. Figure 8 (c) is the THD of phase A voltage. The THD is over the qualified range which needs further Suppression. Figure 8 (d) shows the three-phase voltage unbalance degree. Figure 8 (e) displays the voltage fluctuation value. Figure 8 (f) demonstrates the instantaneous flicker sensation.

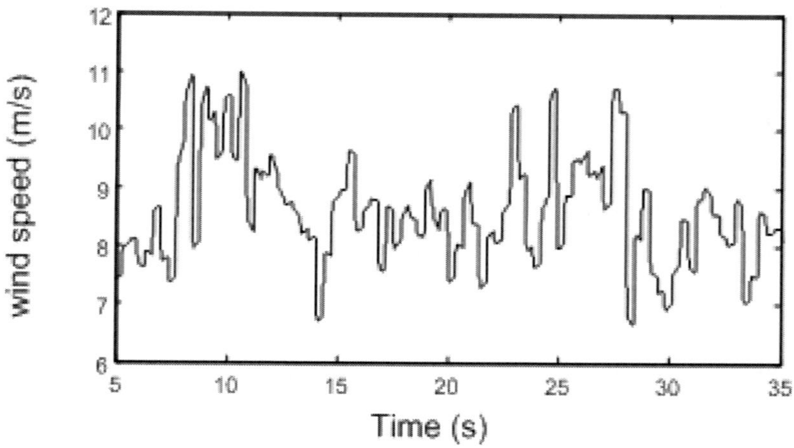

(a)

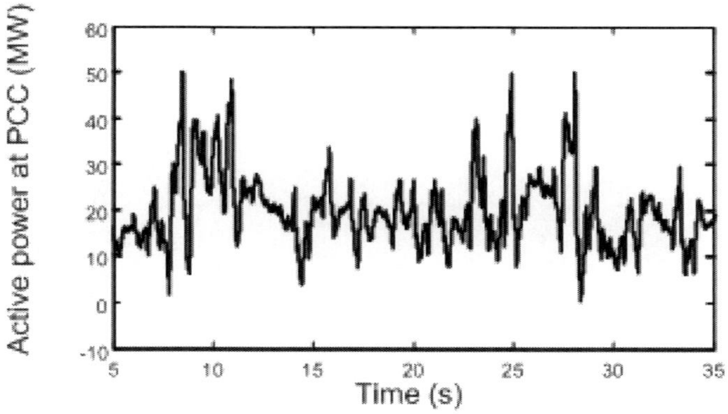

(b)

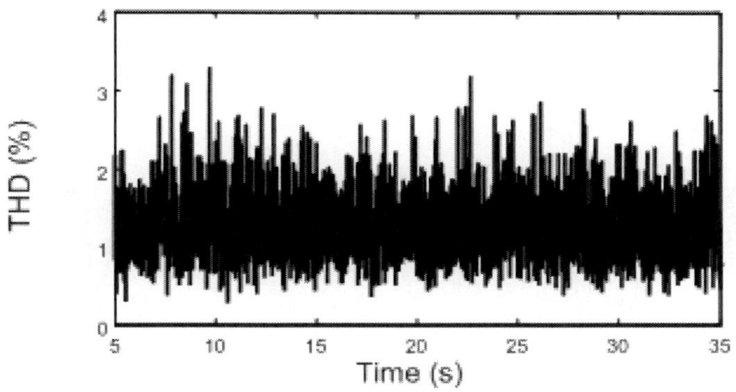

(c)

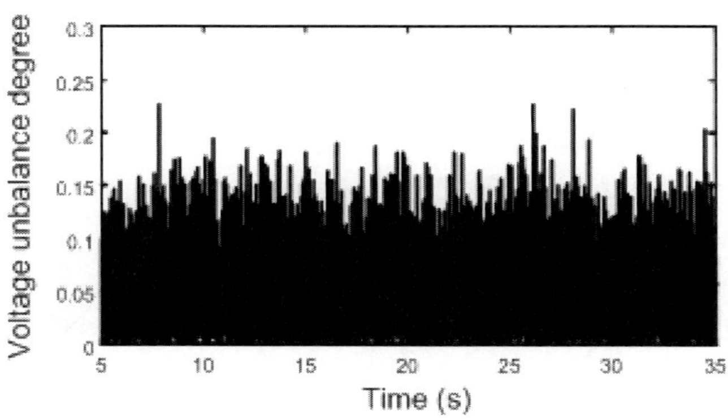

(d)

Figure 8. (Continued).

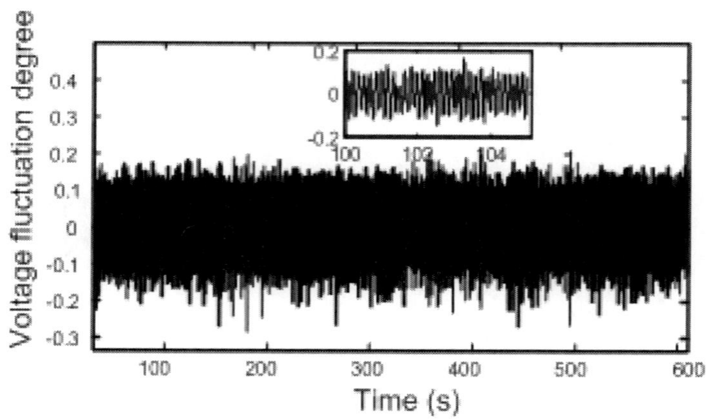

(e)

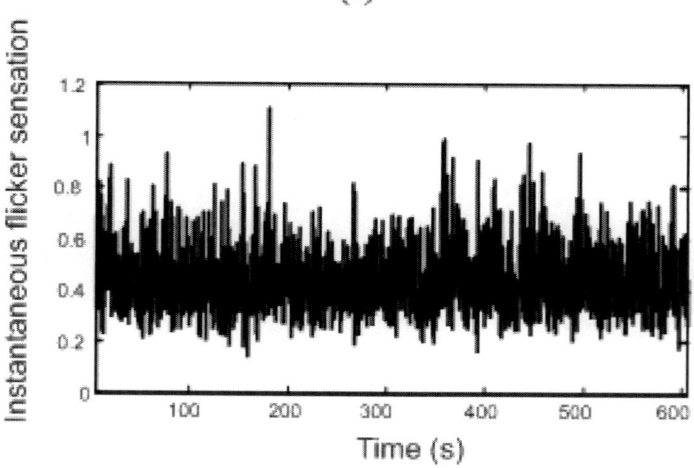

(f)

Figure 8. (a) Wind speed; (b) Active power measured at PCC measured;(c) THD of phase A voltage; (d) three-phase voltage unbalance degree; (e) voltage flicker signal waveform; (f) instantaneous flicker sensation.

Summary: This section established the PSCAD/EMTDC based offshore wind farm model and carried out the power quality analysis at the PCC point of the offshore wind farm integrated into an onshore grid. A set of simulation experiments are carried out to obtain the value of power quality indices, and the numerical result shows that the obtained measurements are within the national limiting value except harmonics.

The measurement and analysis of power quality for grid-connected offshore wind farm is one of the key issues. This section attempts to address this technical challenge through a simulation-based study by the use of PSCAD/EMTDC models. We consider an equivalent model with 50MW capacity for offshore wind farm consisting of 10 doubly fed induction generators (DFIGs) to analyze the power quality at the grid-connected point. A set of indices, including voltage unbalance, harmonics, voltage fluctuation and flicker are discussed. Also, the wavelet transform is adopted to analyze the voltage fluctuation and flicker. Based on the findings and the national power quality standard, we suggest some measures to improve the power quality.

WIND POWER STABILIZATION USING HYBRID ENERGY STORAGE SYSTEMS

The renewable wind power generation has continuously experienced a rapid growth in the past decade driven by the advances of Distributed Energy Resources (DER) and pursuit of low-carbon power generation instead of high dependence of the fossil fuel [27]. However, it should be noted that the output dynamic fluctuation of wind power generation can make a significant adverse impact on both the quality and security of supply in electric power grid. Currently energy storage technology is deemed as a potential solution to address this outstanding issue in wind power generation to cope with low-frequency and high-frequency power fluctuation which can be implemented through mechanical energy storage, electromagnetic energy storage or electrochemical energy storage. More specifically, the HESS includes battery energy storage system (BESS) and super-capacitors energy storage system (SCESS) and various control strategies and algorithmic solutions have been exploited to manage the energy distribution in BESS and SCESS.

The researchers have investigated the method that stabilizing the wind power fluctuations with the help of BESS [27-30, 32-33]. The authors adopted a dual BESS to compensate power mismatch between wind power and desired power schedule for dispatching wind power in [28]. The two BESSs are utilized to cope with the positive and negative power mismatch respectively, and the roles of them can be interchanged according to the predefined thresholds on state of charge (SOC). It is good to extending the lifetime of BESS, but this method has to base on the short-term forecasted wind conditions and cannot deal with the high-frequency part of the wind power fluctuation. In order to solve the high-frequency part problem, HESS have been introduced in [30, 31], in which the super-capacitors energy storage system (SCESS) with a high power density but a lower energy density, is combined with the BESS to achieve a better overall performance. Unfortunately, there has not been an effective way to stabilize the fluctuations in real-time.

In this section, we further address the technical challenge and present a control strategy to manage the output power fluctuation induced by the wind power generation. The key technical contributions made in this work can be summarized as follows: (1) we adopted the low-pass filtering technique to derive the output power stabilization objective which needs to be met through appropriate control strategy; (2) the PSO algorithm is adopted to compute the power of SCESS and BESS whilst considering their constraints to charge/discharge in order to absorb or compensate the wind power through DC/DC transformation. The technical benefits provided by the suggested approach can be in two-fold: One is the control strategy can be implemented in a real-time fashion and without the assumption that the wind power is known as a prior. In the literature, most of the solutions cannot be carried out in real-time and based on the strong assumption that the wind speed can be accurately predicted or estimated. It can predict the wind speed of WF in China inland with the largest error being only 4.46% [35]. Even so, the error between the estimated and the actual wind speed probably exceed the permissible limits; the HESS cannot provide such large power or capacity, that the devices cannot stabilize the fluctuations effectively. Fortunately, this error can be avoided by the method in real-time. The other is that the SCESS is prior to charge or discharge when HESS releases or absorbs power, and the BESS can be considered as the backup equipment which can reduce its usage and extend its lifetime. The charging and discharging of super-capacitors are physical processes, and hence it becomes more economical and environmental friendly.

Hybrid Energy Storage System (HESS)

In this work, we consider that the studied HESS consist of the SCESS as its main component, while BESS works as an auxiliary component. Compared with the energy storage approach based on battery, the super-capacitors have many advantages as follows: (1) Higher power density: The power density of super-capacitors is about 10 to 100 times higher than batteries, and could reach 10kW/kg. This feature makes super-capacitors can be used for scenarios that requires high output power in short time; (2) Faster charging speed: It may only cost dozens of seconds to several minutes to complete the charging process, while the battery needs several hours; (3) Longer lifetime: The cycle lifetime can be over [00000 times of charging, which is 10 to 100 times more than the lifetime of battery; (4) More environmental friendly: The charging and discharging of super-capacitors are physical processes which has limited affections to the environment. Based on the above recognition, we propose an approach of power energy dispatch which works as follows: SCESS has higher priority when power is needed for compensating the power output of wind farm. If the power for compensation is low and changing frequently, SCESS can complete the compensation task independently. However, if the power for compensation is high, the capacity of SCESS cannot be enough for the large power throughput. In this case, BESS works as backup and start to work to compensate the power. This section attempts to propose a type of HESS by using the SCESS as the key component to stabilize the power fluctuation over time of wind farm due to its intermittent nature.

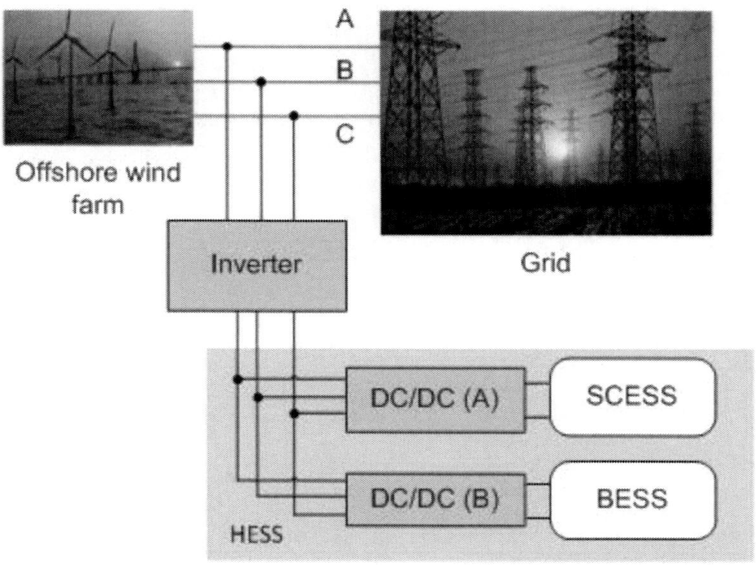

Figure 9. The system structure of HESS.

The HESS system structure is showed in Figure 9. In HESS, the SCESS and BESS are organized through a bus connection. With such structure, once the output power of wind farm is lower than the stabilization target, HESS compensates the error. Otherwise, HESS absorbs the excess power generated from the wind farm. In this way, the wind power can be improved to the standard to be connected to the power distribution grid.

Operating Principle and System Model

One of the key benefits of the proposed approach is that the control strategy can be carried out in a real-time fashion. More specifically, the finding of stabilization target, and charging/discharging process are implemented in real-time. According to the real-time output power, P_{real}, and the stabilization target power, P_{target}, we can easily obtain the power task of HESS P_{S_HESS}.

$$P_{S\_HESS}(t)=P_{target}(t)-P_{real}(t)$$

Through some strategies and algorithms, we can compute the optimal power schedule of SCESS and BESS, $P_{S\_SCESS}(t)$ and $P_{S\_BESS}(t)$ in real-time. DC-DC (A) controls the actual charging or discharging power of SCESS $P_{SCESS}(t)$, and DC-DC (B) controls the actual charging or discharging power of BESS $P_{BESS}(t)$.

In order to control $P_{SCESS}(t)$ and $P_{BESS}(t)$ accurately, we must measure the terminal voltage in real-time. Take SCESS for example, the terminal voltage of SCESS is $U_{SC}(t)$, then the control target of the current is:

$$I_{SC}(t)=\frac{P_{S\_SCESS}(t)}{U_{SC}(t)}$$

Through the close-loop control to the current, we can control the power accurately. The high side of DC-DCs connects to three-phase converter to guarantee the voltage stable.

Stabilization Strategy and Optimization Algorithm Design

A. Stabilize Target Computation

Without a universally accepted standard, we use the theory of low-pass filtering to compute the stabilize target. The high frequency part of wind power is stabilizing by HESS, and the low frequency part is input into the grid. According to the surplus capacity of HESS, we adjust the time constant of the low-pass filter $\tau(t)$ by second. When the surplus capacity is sufficient, we increase $\tau(t)$ appropriately, making the target at time t close to the real-time power at time $t-1$ as much as possible. This method helps to decrease the rate of wind power's change. The stabilize target at time t, i.e., $P_{target}(t)$ and the extra power $\Delta P(t)$ are calculated by (3) and (4).

$$P_{target}(t) = (1 - \lambda(t))P_{stab}(t-1) + \lambda(t)P_{real}(t)$$

$$\Delta P(t) = (1-\lambda(t))(\mathrm{P}_{real}(t) - P_{stab}(t-1))$$

$$P_{stab}(t) = P_{real}(t) + P_{SCESS}(t) + P_{BESS}(t)$$

where $\lambda(t)$ is the filter coefficient at time t, and $\lambda(t) = \Delta t / (\tau(t) + \Delta t)$. $P_{stab}(t)$ is the power after stabilizing.

The relationship between $\lambda(t)$ and surplus capacity is showed as follow.

$$\lambda(t) = \begin{cases} k_1\left(1 - \dfrac{E_{B\_max} - E_B(t-1)}{E_{B\_max} - E_{B\_\min}} \dfrac{E_{SC\_max} - E_{SC}(t-1)}{E_{SC\_max} - E_{SC\_\min}}\right), \Delta P(t) \geq 0 \\ \\ k_2\left(1 - \dfrac{E_B(t-1) - E_{B\_min}}{E_{B\_max} - E_{B\_\min}} \dfrac{E_{SC}(t-1) - E_{SC\_min}}{E_{SC\_max} - E_{SC\_\min}}\right), \Delta P(t) < 0 \end{cases}$$

where k_1 and k_2 are proportionality coefficients. If k_1 or k_2 is relatively small, $\lambda(t)$ is also small, then the stabilize target is relatively smooth, but the power task of HESS is heavy. The harm is that the capacity of HESS maybe not enough and the system cannot work as plan. If k_1 or k_2 is relatively large, $\lambda(t)$ is also large, the stabilize target still has many fluctuations that the grid cannot accept.

B. Power Distribution of HESS

The power task of HESS is described in formula (1), then we should distribute the power task, $P_{S\_SCESS}(t)$ and $P_{S\_BESS}(t)$, based on the state of charge (SOC) of SCESS and BESS. When the system need HESS releases or absorbs power, we use SCESS first. The SOC of SCESS is computed by the following.

$$SOC_{sc}(t) = SOC_{sc}(0) + \int_0^t P_{sc}(\zeta) \cdot d\zeta / E_{sc}$$

$$SOC_{sc}(t) = SOC_{sc}(t-1) + P_{sc}(t)$$

Then we use different strategies in different situations:

Scenario I: If $P_{target}(t) < P_{real}(t)$, the HESS needs to absorb power, then we discuss three cases respectively.

(1) $SOC_{sc}(t-1) < 0.7$, the surplus capacity of SCESS is enough for absorbing power, so the SCESS can complete the power task independently.

$$P_{S\_SCESS}(t) = P_{S\_HESS}(t) = P_{target}(t) - P_{real}(t)$$

$$P_{S\_BESS}(t) = 0$$

(2) $0.7 \le SOC_{sc}(t-1) < 0.9$, the surplus capacity of SCESS is not that enough, then the SCESS should complete the power task cooperating with the BESS.

(3) $SOC_{sc}(t-1) \geq 0.9$, the SCESS stops working in this case. For protecting the devices, we must set a maximum range to the SOC of SCESS.

Scenario II: If $P_{target}(t) > P_{real}(t)$, the HESS needs to releases power, then we also discuss three cases.

(1) $SOC_{sc}(t-1) > 0.4$, the energy in SCESS is enough for releasing power, so the SCESS can complete the power task independently.

$$P_{S\_SCESS}(t) = P_{S\_HESS}(t) = P_{target}(t) - \mathrm{P}_{real}(t)$$

$$P_{S\_BESS}(t) = 0$$

(2) $0.4 \geq SOC_{sc}(t-1) > 0.2$, the energy in SCESS is not that enough, then the SCESS should complete the power task cooperating with the BESS.

(3) $SOC_{sc}(t-1) \leq 0.2$, the SCESS stops working in this case.

C. Algorithm Implementation to Compute the Power Task

In case (1), we use PSO algorithm to compute the optimal power task of SCESS. For making the error between $P_{S\_SCESS}(t)$ and $P_{target}(t)$ small as much as possible, we consider 4 points: $P_{S\_SCESS}(t)$, $P_{S\_SCESS}(t+1)$, $P_{S\_SCESS}(t+2)$ and $P_{S\_SCESS}(t+3)$. We take the maximum rate of power change of SCESS ΔP_{SCESS_MAX} into consideration.

The objective function is:

$$f = \min \sqrt{\frac{1}{4}\sum_{i=0}^{3} \begin{matrix}(P_{S\_SCESS}(t+i) \\ -P_{target}(t+i))^2\end{matrix}}$$

Considering the constraints as follows:

$$P_{S\_SCESS}(t) \leq P_{SCESS\_MAX}$$

$$\left|P_{S\_SCESS}(t+1) - P_{S\_SCESS}(t)\right| \leq \Delta P_{SCESS\_MAX}$$

$$E_{SCESS\_MIN} \leq E_{SCESS}(t) \leq E_{SCESS\_MAX}$$

where P_{SCESS_MAX} is the maximum power of SCESS, E_{SCESS_MAX} and E_{SCESS_MIN} are the maximum and minimum energy in SCESS.

We can get an optimal group of $P_{S\_SCESS}(t)$, $P_{S\_SCESS}(t{+}1)$, $P_{S\_SCESS}(t{+}2)$ and $P_{S\_SCESS}(t{+}3)$, we choose $P_{S\_SCESS}(t)$ as the optimal power task at time t. It could cause a problem that the PSO computing lags the computing of $P_{target}(t)$ by 3 seconds. But it will not affect the real-time system.

Similar method can be adopted to compute the power task of BESS. The difference is we just consider one point, the point at time t, $P_{S\_BESS}(t)$. Because the error between $P_{S\_BESS}(t)$ and $P_{target}(t)$ can be dealt by SCESS.

The objective function is:

$$f = \min \sqrt{\frac{1}{7}\left(\sum_{i=0}^{3} (P_{S\_BESS}(t) - P_{target}(t+i))^2 + \sum_{i=1}^{3} (P_{S\_BESS}(t) - P_{target}(t-i))^2\right)}$$

The constrains are:

$$P_{S\_BESS}(t) \leq P_{BESS\_MAX}$$

$$\left|P_{S\_BESS}(t+1) - P_{S\_BESS}(t)\right| \leq \Delta P_{BESS\_MAX}$$

$$E_{BESS\_MIN} \leq E_{BESS}(t) \leq E_{BESS\_MAX}$$

where P_{BESS_MAX} is the maximum power of SCESS, E_{BESS_MAX} and E_{BESS_MIN} are the maximum and minimum energy in SCESS, ΔP_{BESS_MAX} is maximum rate of power change of BESS.

Numerical Simulation and Numerical Results

This section carries out a set of numerical simulation experiments and discuses the result. We use MATLAB to simulate the effect of the algorithms to compute the stabilize target and distribute the power task of SCESS and BESS. The values of parameters in simulation are listed in Table 2.

A. HESS Stabilization Target Computation

The original output power and the stabilization target are shown together in Figure 10. It can be seen that the stabilization target is more stable than the original output power of wind farm.

Table 2. The Value Of Parameters In Simulation

Rated Power of WF	6 MW
Length of Data	15 minutes
Time Intervals of Data	1 second
Rated Capacity of BESS	0.2 MWh
Max Charge/Discharge Power	2 MW
Max SOC/Min SOC of SCESS	0.9/0.2
Max ramp rate of SCESS	0.12 MW/s
Max ramp rate of BESS	W/s

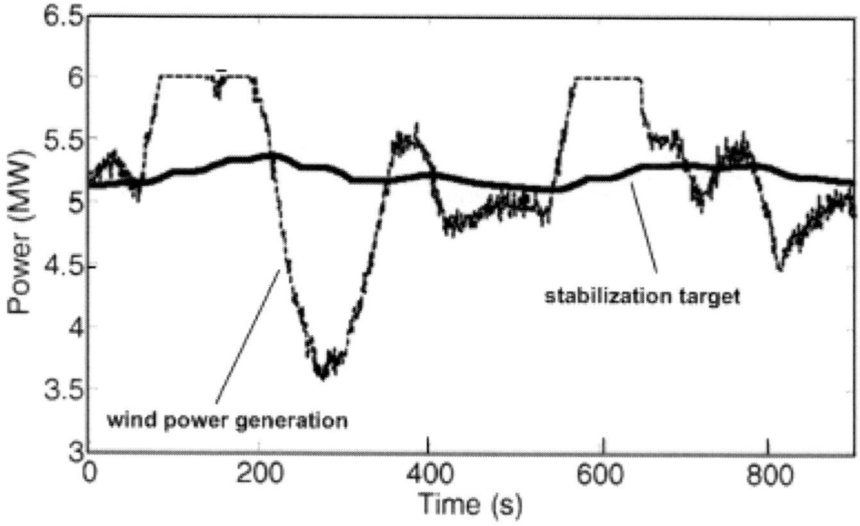

Figure 10. The stabilize target.

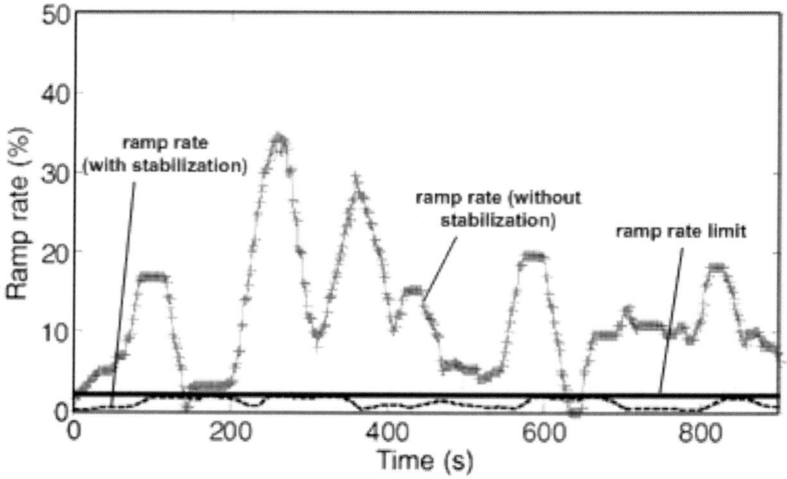

Figure 11. Fluctuation between two adjacent min-points.

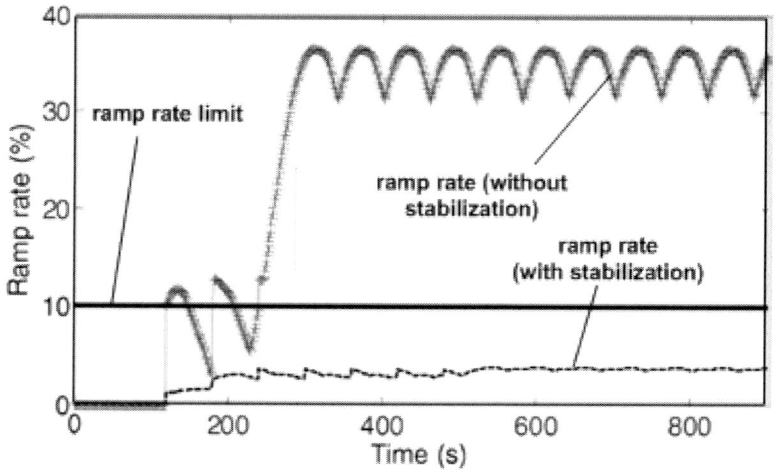

Figure 12. The maximum fluctuation during 30 minutes.

From the result figure, we can find that both the high-frequency fluctuation and the low-frequency fluctuation are reduced.

The power quality reaches the standards that the maximum fluctuation in I minute should be lower than 2% and the maximum fluctuation in 30 minutes should be low than 10%, illustrated in Figure 11 and Figure 12. This improvement of the power quality ensures that the wind power will not broke the stability of the power grid.

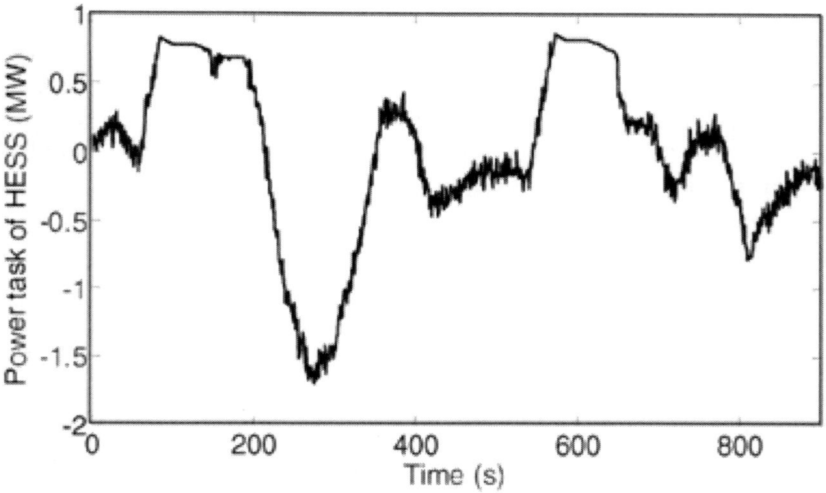

Figure 13. The power task of HESS.

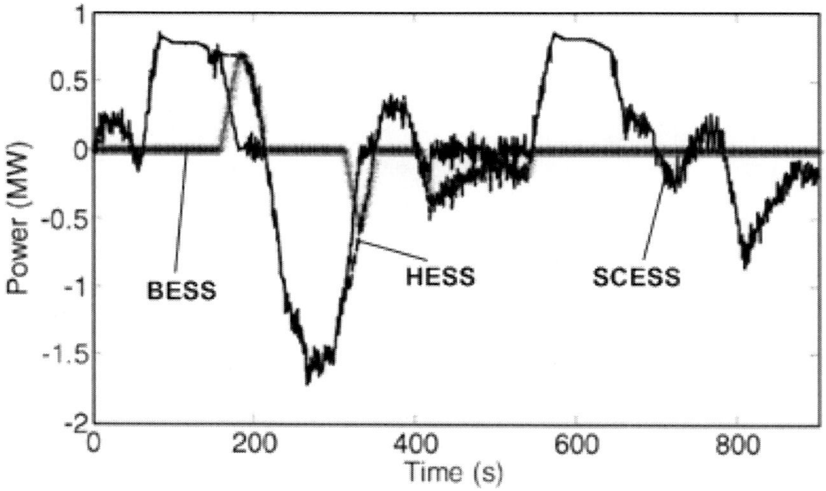

Figure 14. The power target of HESS.

B. Power Distribution of SCESS and BESS

The power task and power garget of HESS are showed in Figure 13 and Figure 14. It can be easily found that the power task fluctuates from -1.7 to 0.8 MW. Also, it includes low-frequency part whose power value is relatively large and high-frequency part which changes fast. This task can be completed by SCESS and BESS with the help of the strategies and algorithms proposed in this chapter. It can be estimated that the capacity and the maximum power of SCESS based on this result.

The strategy of the power distribution changes when the SOC of SCESS exceeds its ideal range, which we chose 0.4 to 0.7 in this simulation. For example, before the moment of 180 s, the power task of SCESS is similar to that of HESS; BESS is not working. After the moment of 180 s, the SOC of SCESS reaches to 0.7, so BESS joins in to complete the task together with SCESS. Then the major power task is completed by BESS and the rate of rise of the SOC of SCESS reduces. This method controls the SOC of SCESS within a relative stable range; and it is helpful to protecting the devices. The power task and the SOC of SCESS are shown together in Figure 15. The SOC increases when SCESS is charging and the SOC decreases when SCESS is discharging. In most of the time, the SOC of SCESS is within the ideal range. However, at the moment of about 300 s, the SOC is below 0.4. In this case, the strategy of the power distribution changes and BESS joins in to work, obviously slowing down the rate of descent of the SOC of SCESS. In addition, from the power task of BESS in Figure 16, we know that the working time of BESS accounts for about 30% of the total time. And if we use a larger capacity SCESS, the BESS will be less used.

The suggested solution of using the BESS as a backup device can decreases scale of BESS and reduce the cost, as well as extend the lifetime of BESS and friendly to the environment.

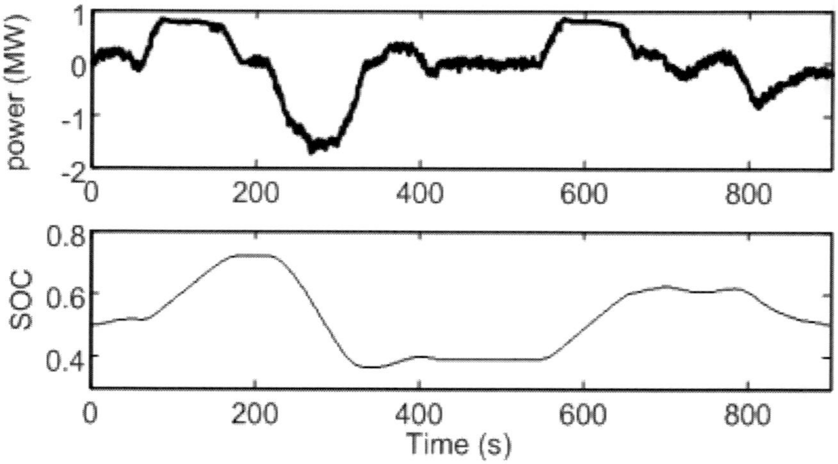

Figure 15. The power task and SOC of SCESS.

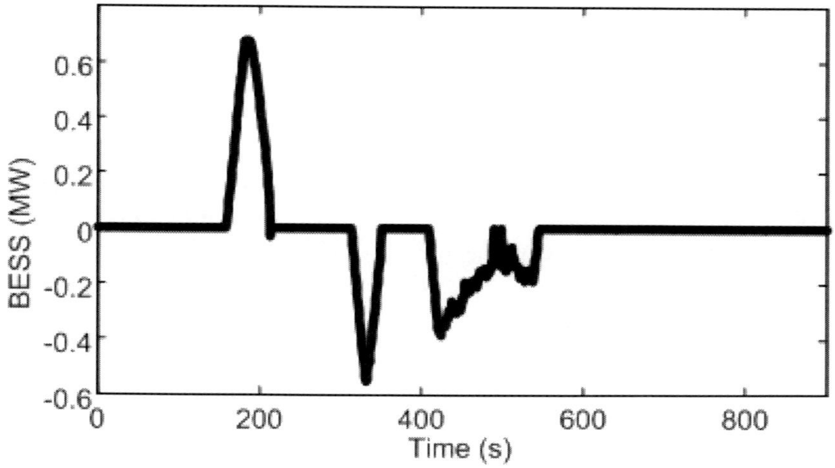

Figure 16. The power task of BESS.

Summary

In recent years, much research effort has been made aiming to alleviate and smooth the power fluctuation induced by the wind farm due to its intermittent nature. The integration of hybrid energy storage system (HESS) including battery energy storage system (BESS) and super-capacitors energy storage system (SCESS) has been considered one of the appropriate solutions to meet the technical challenge of power stabilization in power grid consisting of distributed renewable generations. This section presented a control strategy on managing the HESS to stabilize the power fluctuation in a real-time fashion without the need of predicting wind speed statistics. By the use of low-pass filter with varied time constants to obtain the power stabilization objective, the suggested approach is implemented by distributing the tasks of SCESS and BESS through the particle swarm optimization (PSO) algorithm. The proposed control strategy is assessed through a set of simulation experiments and the result demonstrates its effectiveness in stabilizing the power fluctuation of wind farms.

In our work, the low-pass filter technology with variable time constant is used to compute the stabilize target in real-time. The difference between the target and real output power of WF is the power task of HESS. We utilize PSO algorithm to distribute the power task of SCESS and BESS in real-time, including constrains of devices into consideration. For extending devices' lifetime and decreasing the pollution to the environment, we take the strategy that SCESS is prior to charge and discharge, BESS joins in to work when the capacity of SCESS isn't enough. From the simulations, we can find that SCESS takes up the majority of the power task; the usage rate of BESS is relatively low and there's hardly any transition between charging and discharging of BESS. It not only relieves the devices' burden but also decreases the devices' cost. Moreover, the power distribution methods are different when the SOC of SCESS changes. It can maximize the use of SCESS, and ensure the capacity range. Simulation results demonstrate the feasibility and effectiveness of the proposed real-time strategy in stabilizing the fluctuation of wind power.

CONCLUSION, DISCUSSIONS AND REMARKS

This chapter presented the approaches of establishing the wind farm equivalent model based on PSCAD/EMTDC, and studied the measurements, analysis and improvement approaches of power quality of offshore wind farms. A collection of concluding remarks can be summarized as follows:

This chapter carried out the study of equivalent model and static var compensation for offshore wind farm implemented on PSCAD/EMTDC. In this chapter, the equivalent model of an offshore wind farm under construction and its transmission system is established on PSCAD/EMTDC, including the paralleled transformation based internal electrical network equivalent model and the semi-aggregated equivalent model of wind farm by aggregating the wind turbines with the similar wind speed into a single wind generator, HVAC transmission system and reactive power compensation SVC. The numerical results obtained from simulation experiments demonstrated that the equivalent model of offshore wind farm can represent the collective response of the wind farm effectively.

This chapter further proposed a real-time control strategy to mitigate the wind power fluctuations with HESS. Considering the inherent uncertainty of wind power, IFLF method is proposed to achieve the expected smoothed wind power satisfied the two-time-scale MFPRs, in which filtering time constant is adjusted at one second time scale according to the former fluctuations smoothing results. As for the power sharing between the battery and super-capacitor in HESS, power management strategy based on PSO algorithm is proposed, in which the characteristics of the two kinds of energy storage is utilized while satisfying the battery constraints about SOC and minimal charging/discharging state changing times to prolong the battery lifetime. The simulations demonstrate the effectiveness of our proposed wind power mitigation strategy with HESS. In respect to the future work, a number of directions are considered worth further research efforts.

Based on the model of offshore wind farm and its transmission system, our future work will focus on the further development of more effective power quality improvement scheme. We will enlarge the total capacity of offshore wind farm to 200MW. The active Power Filter detection method will be used to reduce harmonics. Also, we consider that STATCOM can be used to improve the power quality. In addition, we will concentrate on perfecting the discrepancy coefficient and researching more available scheme on power quality assessment. The aim of comprehensive evaluation on power quality is to grasp the global performance of offshore wind power integration. It is meaningful to improve power quality when power quality cannot meet the requirement imposed by the power system.

ACKNOWLEDGMENTS

This work is in part by Zhejiang Province Natural Science Foundation of China (Grant No. LZ15E070001), National High-Tech R&D program of China (863 Program No. 2012AA051704 and 2011AA05A113), and National Science Foundation of China (No. 51107113).

REFERENCES

[1] L. Sun, Z. Q. Mi, Y. Yu, etc. “Active Power and Reactive Power Regulation Capacity Study of DFIG Wind Turbine,” in Sustainable Power Generation and Supply, 20009. SUPERGEN ’09. vol. 33, pp. 1-6, Apr. 2009.

[2] Q. H. Liu, Y. K. He, R. D. Zhao, “Operation and Control of AC-Exited Variable-Speed Constant-Frequency Wind Power Generation System,” *Transactions of China Electrotechnical Society,* vol. 23, pp. 129-136, Jun. 2008.

[3] H. Camblong, I. M. Alegria, M. Rodriguez, G. Abad, “Experimental evaluation of wind turbines maximum power point tracking controllers,” *Energy Conversion and Management*, vol. 47, pp. 2846-2858, Nov. 2006.

[4] E. Muljadi, A., Ellis, J. Hochheimer, etc. "Equivalencing the Collector System of a Large Wind Power Plant," IEEE Power Engineering Society, Annual Conference, Montreal, Quebec, Jun. 2006, pp. 12-16.

[5] L. Q. Liu, Y. Yu, Z. Wang, L. Wang, "Dynamic Equivalence Method of Variable Speed Wind Turbine with Doubly-Fed Induction Generators," Proceedings of the CSU-EPSA, vol. 24, No. 2, pp. 63-66, Apr. 2012.

[6] Y. J. Zhao, Y., Zeng, "Research on applied dynamic equivalence model for wind farm based on Doubly-fed wind power generating units," *Guangdong Electric Power*, vol. 25, No. 4, pp. 53-59, Apr. 2012.

[7] Y. Q. Jin, H. Huang, P. Ju, X. P. Pan, "Collector network transformation methods for wind farm aggregated modeling," *Power System Protection Control*, vol. 40, No. 16, pp. 34-41, Aug. 2012.

[8] M. K. Dosoglu, A. B. Arsoy, "Investigation of various voltage sag analyses in wind farm through SVC," Electrical and Electronics Engineering (ELECO), 7th International Conference, Dec. 2011, pp. I-13-I-17.

[9] A. Qzturk, K. Dosoglu, "Investigation of the control voltage and reactive power in wind farm load bus by STATCOM and SVC," Electrical and Electronics Engineering International Conference, Nov. 2009, pp. I-60-I-64.

[10] L. M. Fernandez, F. Jurado, J. R. Saenz, "Aggregated dynamic model for wind farms with doubly fed induction generator wind turbines," *Renewable Energy*, vol. 33, issue 1, pp. 129-140, Jan. 2008.

[11] M. G. Sugirtha and P. Latha, "Analysis of power quality problems in grid connected wind power plant," in Proc. Conf Recent Advancements in Electrical, Electronics and Control Engineering (ICONRAEeCE), Sivakasi, 2011, pp. 19-24.

[12] J. Yu and 1. Zheng, "Offshore wind development in China and its future with the existing renewable policy," *Energy Policy*, vol. 39, no. 12, pp. 7917-7921, 2011.

[13] S. F. Zyga and K. G. Callister, Offshore Wind Energy in the U.S. Development Strategy and Resources. 2012, pp.20-21.

[14] N. T. Linh, "Power quality investigation of grid connected wind turbines," in Proc. 4th IEEE Conf, industrial Electronic and Applications (lCiEA), Xi'an, 2009, pp. 2218-2222.

[15] X. Y. Lv, et al., "Research on detection method of power quality for wind power connected to grid," in Proc. Conf. 4th Electronic System-integration Technology, Amsterdam, 2012, pp. 1553-1558.

[16] Z. Q. An, et al., "Analysis and testing of wind farm power quality in Inner Mongolia," in Proc. Conf. Electrical and Control Engineering (ICECE), Yichang, 2011, pp. 2002-2005.
[17] C. Wei et aI., "Voltage fluctuation and flicker assessment of a weak system integrated wind farm," in Proc. Conf IEEE Power and Energy Society General Meeting, San Diego, 2011, pp. 1-5.
[18] W. J. Hu, "The impact of Nan'ao wind farm integration on power quality of Shantou Power Grid," in Proc. Conf Electrical Machines and Systems (ICEMS), Wuhan, 2008, pp. 2596 -2601.
[19] E. Vichez and J. Stenzel, "Wind energy integration into 110 kV system - impact on power quality of MV and LV networks," in Proc. IEEE Conf Transmission and Distribution ConJerence and Exposition, Bogota, 2011, pp. 1-6.
[20] A. Abrantes, "Overview of power quality aspects in wind generation," in Proc. Conf. North American Power Symposium (NAPS), Champaign, 2012, pp. 1-6.
[21] IEEE Recommended Practice-Adoption of IEC 61000-4-15:2010, Electromagnetic compatibility (EMC)-Testing and measurement techniques-Flicker meter-Functional and design specifications, IEEE Standard 1453-2011, 2011.
[22] X. Xiao, Analysis and Control of Power Quality. Beijing: China Electric Power Press, 2004, pp. 54-201.
[23] Y. C. Miao, 'The impact of large-scale offshore wind farm on the power system" in Proc. Conf. Electricity Distribution (CiCED), Nanjing, 2010, pp. 1-5.
[24] X. L. W and Y. C, "Measurement of voltage fluctuation and flicker in electric power system based on wavelet transform," in Proc. Conf. Wavelet Analysis and Pattern Recognition (ICWAPR), Beijing, 2007, pp. 1822-1826.
[25] P. K. Ray, et al., "Power quality disturbance detection in grid-connected wind energy system using wavelet and S-transform," in Proc. Conf. Power, Control and Embedded System (ICPCES), Allahabad, 2010, pp. I-4.
[26] R. Dubey, et al., "Detection of power quality disturbances in presence of DFIG wind farm using wavelet transform based on energy function," in Proc. Conf Power and Energy Systems (lCPS), Chennai, 2001, pp. 1-6.
[27] J. Cao et aI., "Battery balancing methods: A comprehensive review," in Proc. IEEE Veh. Power Propulsion Conf, Sep. 2008, pp. 1-6.

[28] Y. Yuan et ai., "Applications of battery energy storage system for wind power dispatch ability purpose," Electric Power Systems Research, 2012, vol. 93, pp. 54- 60.
[29] S. M. Lukic et ai., "Energy storage systems for automotive applications," *IEEE Trans. Ind. Electron.*, vol. 55, pp. 2258-2267, Jun. 2008.
[30] S. M. Lukic et al., "Power management of an ultra-capacitor/battery hybrid energy storage system in an HEY," in Proc. IEEE Veh. Power Propulsion Conf, Windsor, Sep. 2006, pp. 1-6.
[31] J. Chao and A. Emadi. "A new battery/ultra capacitor hybrid energy storage system for electric, hybrid, and plug-in hybrid electric vehicles," *IEEE Transactions on Power Electronics*, vol. 27, Jun. 2012, pp. 122-132.
[32] K. Ogimi et aI., "Optimum operation planning of wind farm using forecasted power data of wind turbine generators considering forecasted error," in Proc. International Conference on Power Electronics - ECCE Asia, May 30-June 3, 2011, pp.1344-1349.
[33] H. Daneshi and A. K. Srivastava, "Impact of battery energy storage on power system with high wind penetration," Transmission and Distribution Conference and Exposition, pp. 1-8,2012 IEEE PES.
[34] W. Zhang and W. M. Wang, "Wind speed forecasting via ensemble Kalman filter," in Proc. *Advanced Computer Control (ICACC),* vol. 2, pp. 73-77, Mar. 2010.
[35] H. W. Guo and H. Y. Dong, "Research on wind speed prediction based on fuzzy neural network," *Eleltric Drive Automation*, vol. 34, pp.I-5, 2012.

BIOGRAPHICAL SKETCH

Name: Qiang Yang

Affiliation: College of Electrical Engineering,
Zhejiang University.

Education: PhD, Queen Mary University of London, UK

Address: 38 Zheda Rd. Zhejiang University,
Hangzhou, Zhejiang PRC

Research and Professional Experience: Qiang Yang held the BS degree (first class honors) in Electrical Engineering and received the M.Sc. (with distinction) and Ph.D. degree both in Electronic Engineering and Computer Science from Queen Mary, University of London, London, U.K., in 2003 and 2007, respectively. He has worked as a Postdoctoral Research Associate at the Department of Electrical and Electronic Engineering, Imperial College London, U.K., from 2007 to 2010 and involved in a number of high-profile U.K. EPSRC and European IST research projects. Currently he is an Associate Professor at College of Electrical Engineering, Zhejiang University, China, and has published more than 100 technical papers and co-authored two books. His research interests over the years include communication networks, smart energy systems, and large-scale complex network modeling, control, optimization, and simulation. He is the member of various international academic bodies including IEEE, IET and IEICE as well as the Senior Member of China Computer Federation (CCF).

Professional Appointments: Associate Professor, Electrical Engineering, Zhejiang University, PRC

Honors:

Publications Last 3 Years:

Book Chapters

1. Q. Yang, Z. Bao, W. Yan, T. Wu, Book Chapter "Smart energy management in microgrid with wind power generators and plug-in electric vehicles," Book Ed. *Plug-in Electric Vehicles in Smart Grid: Management and Control Strategies*, Sumedha Rajakaruna, Farhad Shahnia, Arindam, Ghosh (Eds.) Springer, Dec. 2014.

Referred Papers

1) X. Fang, S. Ma, Q. Yang, J. Zhang, "Cooperative energy dispatch for multiple autonomous microgrids with distributed renewable sources and storages," *Energy*, in press, Jan. 2016. (IF=5.153).
2) X. Fang, Q. Yang, J. Wang, W. Yan, "Coordinated dispatch in multiple cooperative autonomous islanded microgrids," *Applied Energy*, vol. 162, pp. 40-48. Jan. 2016. (IF=6.330).

3) Y. Zhang, W. Yan, Q. Yang, "Network-based leader-following consensus for second-order multi-agent systems with nonlinear dynamics," *Transactions of the Institute of Measurement and Control*, pp. 1-9, April, 2015.
4) B. Ruan, Q. Yang, W. Yan, "Demand response under real-time pricing for domestic energy systems," *Journal of Mechanical and Electrical Engineering*, April. 2015.
5) R. Dong, Q. Yang, W. Yan, "Dynamic optimal network topology reconfiguration based on multi-agent system approach," *Journal of Zhejiang University, S*ept., 2015.
6) Z. Bao, Q. Zhou, Z. Yang, Q. Yang, T. Wu, "A multi time-scale and multi energy-type coordinated microgrid schedule solution- Part I: Model and methodology," *IEEE Trans. on Power systems*, vol. 99, Sept. 2015. (IF=3.889).
7) Z. Bao, Q. Zhou, Z. Yang, Q. Yang, T. Wu, "A multi time-scale and multi energy-type coordinated microgrid schedule solution- Part II: Optimization algorithm and case studies," *IEEE Trans. on Power systems*, vol. 99, Sept. 2015. (IF=3.889).
8) Q. Yang, Q. Yang, W. Yan, T. Wang, "PSO- based optimized LS-SVM approach for fault prediction of primary air fan," in Proc. Chinese Automation Congress (CAC), Wuhan, China, 2015.
9) Q. Wang, Q. Yang, W. Yan, "Optimal dispatch in residential community with DGs and storage under real-time pricing," in Proc. IEEE International Conf. Information and Automation, Yunnan, 2015.
10) L. Ye, Q. Yang, W. Yan, "A scalable coordinated approach towards optimal reactive power compensation in distributed network with renewable DGs," in Proc. Chinese Control Conference (CCC), Hangzhou, 2015.
11) J. Yao, Q. Yang, W. Yan, "Scalable DG-based power supply restoration in distribution network under blackouts," in Proc. International Conference on Environmental Science and Material Application, 2015.
12) Y. Lu, Q. Yang, W. Xu, Z. Lin, W. Yan, "Cyber security assessment in PMU-based state estimation of smart electric transmission networks," in Proc. 27th Chinese Control and Decision Conference (CCDC), Qingdao, 2015.
13) H. Zeng, Q. Yang, "Adaptive voltage regulation of islanding DC microgrid with multiple distributed PVs and storage units," in Proc.

27th Chinese Control and Decision Conference (CCDC), Qingdao, 2015.

14) Y. Zhang, W. Yan, Q. Yang, "Individual synchronization control of complex network with non-identical nodes," in Proc. 27th Chinese Control and Decision Conference (CCDC), Qingdao, 2015.
15) H. Tang, J. Zheng, Q. Yang, "Current status and key technology of protection system for DC microgrids," *Electrical Apparatus,* (in Chinese), pp. 1-6, no.10, 2014.
16) Y. Qian, C. Wu, Q. Yang, "On the topological characterization of sharing video network: Youtube as a case study," *Chinese Journal of Electronics*, in press, 2014.
17) Y. Zhang, P. Li, Q. Yang, W. Yan, "Synchronization criterion and control scheme for Lur'e type complex dynamic networks with switching topology and coupling time-varying delay" *Asian Journal of Control*, vol. 12, Dec. 2014.
18) X. Fang, Q. Yang, W. Yan, "Outer synchronization between complex networks with nonlinear time-delay characteristics and time-varying topological structures," *Journal of Mathematical Problems in Engineering, Special Issue: System Simulation and Control in Engineering*, Elsevier, July, 2014.
19) Y. Zhang, W. Yan, Q. Yang, "Synchronization control of complex dynamic network with non-identical nodes and time-varying topological structures," *Journal of Mathematical Problems in Engineering, Spec issue: Modeling and Control of Complex Networked Systems*, Elsevier, pp. 1-8, July 2014.
20) B. Wang, C. Wu, Q. Yang, et al., "A secure routing model based on distance vector routing algorithm," *Journal of Science China Info. Sci.,* vol. 57, no. 1, Springer-Verlag, Jan. 2014.
21) X. Fang, Q. Yang, W. Yan, "Modeling and analysis of cascading failures in directed networks," *Safety Science*, vol. 65, pp. 1-9, Elsevier, June, 2014 (IF=2.210).
22) X. Fang, Q. Yang, W. Yan, "Outer synchronization for a class of generalized nonlinear delay networks," in Proc. IEEE International Conf. Information and Automation, July 27-30, Innermongolia, 2014.
23) R Li, et al., Q. Yang, "A cost-effective approach for feeder fault detection of radial distribution networks," in Proc. IEEE Transportation Electriflcation Conference and Expo Asia - Pacific, to appear 2014.

24) B. Ruan, Q. Yang, X. Fang, W. Yan, "Demand response under real-time pricing for domestic energy system with DGs," in Proc. IEEE Int'l Conf. on Power System Technology (PowerCon'04), Chengdu, Oct. 2014.
25) H. Zeng, H. Zhao, Q. Yang, "Coordinated energy management in autonomous hybrid AC/DC microgrids," in Proc. IEEE Int'l Conf. on Power System Technology (PowerCon'04), Chengdu, Oct. 2014.
26) C. Ma, Q. Yang, W. Yan, "On the hierarchical control framework for distributed energy storage management in large-scale distribution networks," in Proc. IEEE Int'l Conf. on Power System Technology (PowerCon'04), Chengdu, Oct. 2014.
27) B. Zhou, W. Gao, S. Zhao, X. Lu, Z. Du, C. Wu, Q. Yang, "'Virtual network mapping for multi-domain data plane in software-defined networks," International Conference on Wireless Communications, Vehicular Technology, Information Theory and Aerospace and Electronic Systems (WVITAE), Aalborg, Denmark, May, 2014.
28) R. Dong, Q. Yang, W. Yan, "A two-stage approach on island partitioning of power distribution networks with distributed generation," Proc. the 26th Chinese Control and Decision Conference (CCDC), to appear, 2014.
29) J. Zhang, Q. Yang, W. Yan, "Decomposing furnace temperature modeling approach based on model parameter aggregation," *Journal of Zhejiang University*, Oct. 2013.
30) Y. Miao, Q. Yang, C. Wu, M. Jiang, J. Chen, "Multicast virtual network mapping for supporting multiple description coding based video applications," *Computer Networks*, vol. 57, no. 4, pp. 990-1002, Elsevier, March, 2013.
31) T. Wu, Q. Yang, Z. Bao, W. Yan, "Coordinated energy dispatching in microgrid with wind power generation and plug-in electric vehicles," *IEEE Transactions on Smart Grid*, vol. 4, no. 3, pp. 1453-1463, June 2013 (IF=5.719).
32) Y. Che, Q. Yang, C. Wu, "Towards a hierarchical global naming framework in network virtualization," *SKII Transactions on Internet and Information Systems*, vol. 7, no. 5, pp. 1198-1212, May, 2013.
33) Y. Che, K. Chiow, X. Hong, Q. Yang, "EDA: An enhanced dual-active algorithm for location privacy preserving in P2P mobile networks," *Journal of Zhejiang University -C*, April 2013.

34) C. Li, Q. Yang, W. Yan, "Power supply management with photovoltaic source under failure conditions," *Journal of Power System Protection and Control*, vol. 8, 2013.
35) X. Fang, Q. Yang, W. Yan, "Topological characterization and modeling of dynamic evolving power distribution networks," *Simulation Modeling Practice and Theory*, Elsevier, vol. 31, pp. 186-196, Feb. 2013.
36) M. Zhang, C. Wu, M. Jiang, Q. Yang, "Robust dynamic virtual network provisioning," *Journal of Chinese Electronics*, vol. 22, no. 1, Jan. 2013.
37) G. Chen, Q. Yang, Z. Bao, W. Yan, "Real-time wind power stabilization approach based on hybrid energy storage systems," Proc. International Conference on Advanced Computational Intelligence, Hangzhou, 2013.
38) T. Zhang, Z. Bao, G. Chen, Q. Yang, W. Yan, "Control strategy for a hybrid energy storage system to mitigate wind power fluctuations" Proc. International Conference on Advanced Computaional Intelligence, Hangzhou, 2013.
39) R. Sun, W. Yan, Q. Yang, Z. Bao, J. Zhang, "Power quality assessment of offshore wind farms based on PSCAD/EMTDC models," Proc. International Conference on Advanced Computational Intelligence, Hangzhou, June 2013.
40) J. Zhang, W. Yan, Q. Yang, Z. Bao, R. Sun, "Power quality measurement and analysis of offshore wind farms," Proc. International Conference on Advanced Computational Intelligence, Hangzhou, June 2013.
41) G. Chen, Z. Bao, Q. Yang, W. Yan, "Scheduling strategy of hybrid energy storage system for smoothing the output power of wind farm," Proc. IEEE International Conference on Control and Automation (IEEE ICCA'13), Hangzhou, PRC. June 2013.
42) F. Zhao, Z. Bao, Q. Yang, W. Yan, "Equivalent model and static var compensation for offshore wind farm implemented on PSCAD/EMTDC," Proc. IEEE International Conference on Control and Automation (IEEE ICCA'13), Hangzhou, PRC. June 2013.
43) G. Shen, Q. Yang, W. Yan, "Adaptive control approach for PMSM based on parameter identification: a simulation perspective," Proc. IET International Conference on Information Science and Control Engineering, 2013.

In: Wind Farms
Editor: Marian Dunn

ISBN: 978-1-63484-841-1

Chapter 4

IMPACT ASSESSMENT OF WIND FARMS ON RADIO DEVICES IN CIVIL AVIATION

Xiaoliang Wang*[*], *Renbiao Wu*, *Weikun He* and *Yuzhao Ma

Tianjin Key Laboratory for Advanced Signal Processing,
Civil Aviation University of China, Tianjin, P. R. China

ABSTRACT

Wind power is an attractive clean energy and wind farms increase with very high speed in recent years. However, as a particular obstacle, wind farms may degrade the performance of radio devices in civil aviation obviously. Therefore, wind farms may threaten the flight safety and correct impact assessment of wind farms on radio devices is important to guarantee the safety of civil aviation. In this chapter the potential impact of wind farms on radio devices in civil aviation and a review of the impact assessment procedure and methods of our research group is presented. The radio devices discussed in the chapter include surveillance devices such as primary surveillance radar (PSR) and second surveillance radar (SSR) and radio navigation devices such as very high frequency omnidirectional range (VOR) and instrument landing system (ILS). A wind farm usually comprises several wind turbines with very large size. The proper estimation of the scattering coefficient or radar cross section (RCS) of the wind turbine is of great importance to assess

[*] Corresponding author: E- mail: wxl_ee@126.com.

the impact of wind farms correctly. However, the intensity of electromagnetic scattering and the RCS of a wind turbine vary with several factors. Consequently a review of RCS estimation methods for a wind turbine of our research group is also presented in this chapter.

Keywords: civil aviation, wind farm, wind turbine, impact assessment, RCS estimation, radio device, radar, radio navigation

INTRODUCTION

Wind power is an attractive clean energy and wind turbines in wind farms grow in numbers and in size with very high speed in recent years. The global cumulative installed wind capacity has increased exponentially in the past ten years. However, some researches reveal that wind farms, as a particular obstacle, may degrade the performance of radio devices in civil aviation obviously. Therefore, wind farms may threaten the flight safety of civil aviation.

The impacts of wind farms on civil aviation have attracted more attention of several countries, organizations and researchers. The Civil Aviation Authority (CAA) of the United Kingdom has published a document named "CAA Policy and Guidelines on Wind Turbines" [1]. In this document, the potential impacts of wind farms on aviation are summarized. These potential impacts not only include the effects of wind farms on radio devices in aviation, but also contain other effects. For example, anemometer masts in a wind farm are often difficult to acquire visually for pilots and they increase the collision risk for aircrafts. Certainly the main impacts are interference of wind farms on radio devices in aviation.

The correct impact assessment of wind farms on radio devices is important for the site selection of a radio device or a wind farm when they are near each other, and is important to guarantee the safety of civil aviation. The air traffic control (ATC) radar is a radio device which is the most possible to be interfered with by wind farms. As a result Eurocontrol and organizations of the United States and Canada have presented their guidelines respectively on the assessment of the potential impact of wind farms on radars [2-4]. Some researchers also have presented their assessment methods for potential impact of wind turbines on radars in some literatures [5, 6]. A research group from Spain presented their assessment method on several telecommmunication services include radars, aeronautical navigation systems, radiolinks and analog and digital terrestrial broadcasting services and developed an assessment software called Wi^2 [6, 7].

For the problem of assessing the impact of wind farms on radio devices in civil aviation, this chapter mainly presents a review of the work of our research group on this subject. Some works of us are based on existed methods of other researchers and some works are original.

The organization of this chapter is as follows. Firstly the potential impacts of wind farms on radio devices in civil aviation are described. Then our impact assessment methods on radio devices are presented. The estimation of radar cross section (RCS) of a wind turbine is very important for the quantitative detailed assessment, so our RCS estimation methods for a wind turbine are also presented. Finally the conclusion of the chapter is given.

IMPACTS OF WIND FARMS ON RADIO DEVICES

A wind farm usually consists of dozens of or more than a hundred wind turbines. A wind turbine is commonly made up of three main parts: tower, nacelle and rotated blades. The height of a wind turbine and the diameter of rotated blades are usually both near or more than 100 meters. The V-164 8MW wind turbine of Vestas has a huge rotor diameter of 164 meter and its total height could be 220 meters. The huge tower of a wind turbine is commonly made of steel and could be a huge scatterer for electromagnetic wave. Although the material of blades of a wind turbine is usually made of complex material of glass fiber and epoxy resin, it makes the blades be an important scatterer for electromagnetic wave that the size of the blades is huge and some metal materials which are used for lightning protection are contained in blades. The maximum radar cross section (RCS) of a wind turbine is usually 20 – 60 dBsm [8, 9].

The wind farm is a particular obstacle. There are several reasons that the interference of the wind farm on radio devices is more difficult to be mitigated than other common obstacles. First of all, the wind turbine has a huge geometrical size and a large RCS, which is larger than that of a Boing 747 aircraft [8]. Secondly, the rotation of the blades of the wind turbine may produce more complex interference on radio devices. The linear velocity of the different parts in a rotated blade varies continuously from the blade hub to the blade tip. It produces wide Doppler frequency spectrum, which brings difficulties to radar utilizing targets' Doppler frequency spectrum such as the air traffic control primary surveillance radar (PSR). In addition, the rotated blades also produce variational multipath scattering for electromagnetic wave. Thirdly, the state of the wind turbine is changeful. The blades' rotation speed, the direction of the blades' rotation plane and the blades' tilt angle will vary with wind direction and wind speed. Therefore, the interference of the wind farm is time-varying and it is difficult to suppress time-varying interference with traditional method. Finally, there are a larger number of wind turbines in a wind farm commonly. It will make an obvious shadow space behind the wind turbines. Also, the scattering of electromagnetic wave between different wind turbines further increases the difficulty to suppress the interference of the wind turbines [10].

Wind farms may impact on several different kinds of radio devices in civil aviation. These devices mainly include aircraft surveillance devices such as primary surveillance radar and secondary surveillance radar (SSR), and radio navigation devices such as very high frequency omnidirectional radio range (VOR) and instrument landing system (ILS).

Impacts on Primary Surveillance Radar

The primary surveillance radar acquires the distance and the azimuth angle of an aircraft through transmitting a pulse electromagnetic wave and receiving the backscattered signal from the aircraft. Currently PSR usually utilizes the moving target detection (MTD) technique to detect the aircraft and to suppress the static clutter.

The wind farm may degrade the performance of PSR in several aspects. American researchers have validated the wind farms' obvious impacts on PSR with fight experiments [11].

Firstly, the wind farm may reduce the probability of detection of the PSR. The wind turbine may shadow the electromagnetic wave transmitted from the radar. The impact of shadow of one wind turbine is limited, but the shadow of dozens of or more than one hundred wind turbines is noticeable. If an aircraft flies in the shadow of a wind farm, it may not be detected by radar. Furthermore, current MTD technique cannot suppress the clutter from the wind turbine because of the rotated blades. In addition, the constant false alarm rate (CFAR) detection is widely used in PSR. The clutter from the wind farm would raise the detection threshold for the range-azimuth bin near wind turbines. Then the probability of detection for the region near wind turbines would reduce.

The second potential impact of wind farm on PSR is it may produce false targets. Some false targets are produced by direct clutter of the wind turbine and some false targets are produced by multipath scattering of wind turbines for the transmitted signal from the radar and the scattered signal from the aircraft.

Processor overload is also an impact of wind farm on PSR. It appears when the number of false targets produced by wind turbines is too high.

Another potential impact of wind farm is receiver saturation. This impact occurs when the power reflected from the wind turbine is too large and beyond the dynamic range of the radar receiver.

In addition, the wind farm may bring some range and azimuth errors for the position of the aircraft. If the path difference between the direct signal and the multipath signal scattered by a wind turbine is small, the direct signal and the multipath signal would overlay and then the position determined by radar will have some errors.

Impacts on Secondary Surveillance Radar

The secondary surveillance radar get the position and the identification code of an aircraft through transmitting the interrogation signal and receiving the response signal from the transponder equipped on the aircraft. The secondary surveillance radar usually has three work modes (A mode, C mode and S mode) and could acquire more information about an aircraft. The PSR usually could get range and azimuth angle of an aircraft only. Besides range and azimuth angle, SSR could also get identification code of an aircraft in mode A and altitude of an aircraft in mode C. In addition, SSR could be used as a data link in mode S. The principle of acquiring the range and azimuth angle for SSR is similar to PSR, but SSR gets other information like a communication device. Because SSR could get whole 3D position information and identification code, SSR is the main surveillance device in civil aviation. Although SSR relies on the response of the aircraft and could be used for cooperating target only, this condition is easy to be satisfied in civil aviation.

Some of the impacts of wind farm on SSR is similar to PSR, as well as some of the impacts are different. American researchers also found the impacts of wind farms on SSR with fight experiments [11].

The wind farm also may reduce the probability of detection of the SSR. One reason of this impact is the same to PSR. It is the shadow effects of the wind farm. Another reason is the multipath scattering for response signal and interrogation signal of SSR, which is different from PSR. For one thing, if the multipath response signal overlaps the direct response signal and some pulses of these two signals overlap, the waveform of the pulses may be distorted and could not be recognized by SSR and then the target would be missed. For another, SSR utilizes P2 pulse, which is 9dB lower at least than P1 pulse in the main lobe to suppress the interrogation from the side lobe. If the P1 pulse of the multipath interrogation signal scattered by the wind farm overlaps the P2 pulse of the direct interrogation signal, the interrogation from the main lobe may be regarded as a side lobe interrogation. Then the aircraft does not response and would be missed in radar.

The multipath scattering for response signal could also produce false targets for SSR. If the false target is very near to the real target, track branching off may occur.

The processor overload may also occur if there are too many false targets.

In addition, if the multipath response signal overlaps the direct response signal, decoding error of altitude and identification or position error may occur for SSR.

It is suggested to treat the wind farm as a static obstacle because SSR does not employ technique like MTD in a technical report of National Telecommunication and Information Administration of the United States [3]. In fact, the rotated blades of a wind turbine may produce time-varying multipath effects. Fortunately the variety of the path difference between the direct signal and the multipath signal caused by time-varying multipath effects is small.

Impacts on VHF Omnidirectional Radio Range

VOR gives a directional guidance through electromagnetic wave in VHF band for pilots. The pilots could get the azimuth angle of the VOR ground station corresponding to the aircraft.

The multipath scattering of the wind farm may make the VOR receiver equipped in the aircraft get some errors of azimuth angle. Some researchers from the United States analyzed the impact of a wind farm on the VOR serving the Sidney Municipal Airport (SNY) at Sidney, Nebraska. The analyzed wind farm contains 40 wind turbines and the distances from the VOR to wind turbines are from 12.5 to 19.5 kilometers. The analysis results indicate that the maximum azimuth angle error caused by the wind farm is 3.1 degree [12]. Another similar analysis is finished by Germany researchers. They consider a case that a wind turbine is 1.4 kilometers away from the VOR. The result is the maximum azimuth angle error is 10 degree for traditional VOR and is 0.72 degree for Doppler VOR (DVOR) in a 1000 meters height plane and in a 100 km x 100 km region.

And what is worse, the azimuth angle error for VOR receiver is not steady and always varies because of the time-varying multipath effects. A pilot would feel confused when he sees a pendular pointer.

Impacts on Instrument Landing System

The instrument landing system provides radio landing guidance for aircrafts. The instrument landing system consists of three different parts: localizer, glide slope and marker beacon. The localizer and the glide slope may be impacted by a wind farm. The multipath effect caused by wind farms also produces the angle measurement error for airborne localizer and glide slope equipments. Therefore, the wind farms increase the danger of aircraft landing.

IMPACTS ASSESSMENT

There are some guidelines, criteria or schemes proposed by different organizations or researchers to assess the impacts of wind farms on radars or other radio devices already [2-6]. Some of the assessment methods in different guidelines, criteria or schemes are similar, but others are different. In this section we mainly introduce our assessment procedure and methods for assessing the impacts of wind farms on radio devices in civil aviation including PSR, SSR, VOR and ILS.

Assessment Procedure

The impact assessment is implemented for different radio devices respectively. For each device a four steps assessment procedure is employed. As is shown in Figure 1, the four steps consist of the geographic position relationship calculation, the zone partition assessment, the visibility assessment and the detailed engineering assessment [10].

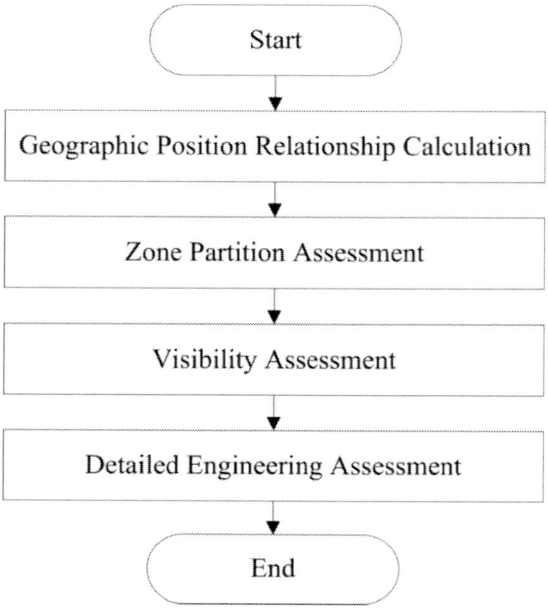

Figure 1. Flow Chart of the Assessment Procedure.

In the geographic position relationship calculation, the distance, azimuth angle and elevation angle from the assessed equipment to each wind turbine in wind farms should be calculated. In the zone partition assessment the space around the assessed equipment is partitioned into different types of zones including safeguarding zones, detailed assessment zones, further assessment zones and no assessment zones. In the visibility assessment the visibility of each wind turbine in the further assessment zones corresponding to the assessed radio device is analyzed. And in the detailed engineering assessment different quantitative calculation for different types of devices is implemented.

Geographic Position Relationship Calculation

The positions of the radio devices and wind turbines in a wind farm are usually provided by longitude, latitude and altitude in the impact assessment. In order to calculate the distance, azimuth angle and elevation angle from the assessed equipment to each wind turbine in a wind farm we employ the method of inverse solution of geodetic problem with the spherical earth model. That is

$$\lambda = \lambda_2 - \lambda_1 \tag{1}$$

$$d = R \cdot \arccos[\sin\phi_1 \sin\phi_2 + \cos\phi_1 \cos\phi_2 \cos\lambda] \tag{2}$$

$$\alpha = \arctan\left[\frac{\sin\lambda \cos\phi_2}{\cos\phi_1 \sin\phi_2 - \sin\phi_1 \cos\phi_2 \cos\lambda}\right] \tag{3}$$

where λ_1 is the longitude of the radio device, and ϕ_1 is the latitude of the radio device as well as λ_2, ϕ_2 are the longitude and the latitude of a wind turbine respectively. d is the horizontal distance from the radio device to a wind turbine and R is the radius of the spherical earth model (the value 6370 km could be used) in formula (2). In formula (3) α is the azimuth angle of the wind turbine corresponding to the analyzed radio device. The exact calculation should utilize the elliptical earth model rather than the spherical earth model, but the spherical earth model is enough for the calculation of the electromagnetic propagation.

The height of a wind turbine is very large. Therefore, it is needed that the elevation angle corresponding to the top and the bottom of the wind turbine are both calculated. The origin for calculation the elevation angle is the center of the antenna of the radio device. Then the elevation angle corresponding to the top and the bottom θ_t, θ_b could be calculated with [10]

$$\theta_t = \arctan\frac{H_w + h_w \quad - \quad H_e + h_e}{d} \tag{4}$$

$$\theta_b = \arctan\frac{H_w \; - \quad H_e + h_e}{d} \tag{5}$$

where H_w, H_e is the altitude above the sea level of the wind turbine and the radio device respectively and h_w is the maximum height of the wind turbine as well as h_e is the height of the center of antenna of the radio device corresponding to the ground. d is the horizontal distance from the radio device to the wind turbine, which is the same as that in formula (2).

Zone Partition Assessment

The zone partition criteria are different for different radio device.

For PSR, we utilize the zone partition criterion similar to the impact assessment guidelines of Eurocontrol [2]. The space where the horizontal distance to the PSR is less than 500 meters is safeguarding zone; the space where the horizontal distance to the PSR is from 500 meters to 15 kilometers is detailed assessment zone and the space where the horizontal distance to the PSR is larger than 15 kilometers is further assessment zone.

For SSR, we also employ the zone partition criterion similar to the impact assessment guidelines of Eurocontrol [2]. The space where the horizontal distance to the SSR is less than 500 meters is safeguarding zone; the space where the horizontal distance to the SSR is from 500 meters to 40 kilometers is further assessment zone and the space where the horizontal distance to the PSR is larger than 40 kilometers is no assessment zone. The distance from the SSR to the boundary of the no assessment zone is 16 kilometers in Eurocontrol guideline. 16 kilometers is calculated with the transmitted power of 2000 W and the wind turbine's RCS of 35 dBsm. However, wind turbine's RCS may reach 60 dBsm and the transmitted power may be larger than 2000 W. Thus, we use 40 kilometers which is calculated with the transmitted power of 4000 W and the wind turbine's RCS of 60 dBsm for more cautious assessment.

For VOR, the localizer and the glide slope of ILS, we utilize "European guidance material on managing building restricted areas" [14]. The space inside the building restricted areas is the further assessment zone and the space outside the building restricted areas is no assessment zone.

Visibility Assessment

In further assessment zones, we further analyze the visibility of each wind turbine corresponding to the assessed radio device. We presented a visibility analysis method in the reference [10].

We employ the digital elevation model (DEM) to construct a terrain profile from the assessed radio device to a wind turbine. The first Fresnel zone is utilized to consider the nonlinear propagation of the electromagnetic wave like the reference [3].

There are some public world terrain data resource such as SRTM data of 90 m grid or 30 m grid and ASTER GDEM data of 30m grid. The interpolation of DEM data may be required for some visibility assessment cases. Because the SRTM data and ASTER GDEM data are both produced by remote sensing, the accuracy of data is limit. In addition, the terrain may be level off during the construction of wind farms or radio devices. Therefore, if there are some more accurate terrain data such as the altitude of the base of the wind turbine, the DEM should be amended with the more accurate terrain data.

The first Fresnel zone is based on the line connecting the antenna center of the assessed radio device and the top or the bottom of the wind turbine. Some visibility analysis results are shown in Figure 2. As is shown in Figure 2, the first Fresnel zone is a spindle shape space. In the position where the slant distance to the two endpoints of the spindle are d_1 and d_2 respectively, the radius of the section could be calculated with

$$R_F = \sqrt{\frac{\lambda}{1/d_1 + 1/d_2}} \tag{6}$$

where λ is the wavelength of the electromagnetic wave.

The visibility analysis devides wind turbines in further assessment zones into three groups: visible completely, visible partly and invisible completely. Visible completely and visible partly wind turbines should be further assessed in detail like wind turbines in detailed assessment zones, and invisible completely wind turbines won't make obvious interference on the radio device.

The visibility analysis above only analyzes the visibility along the tower of the wind turbine. However, the wind turbine has the rotated blades of large diameter. Therefore, for the visible completely or invisible completely wind turbines, some further visibility analysis may be required. For this fact, we proposed a detailed visibility analysis method for the wind turbine in the reference [15]. The direction of the blades' rotation plane varies with wind direction, so we model the rotated blades with a sphere in this method. The tower of the wind turbine is a truncated cone and could be modeled as a thin cylinder in this method. Then the visibility is analyzed point by point in a specific section of the sphere modeled by rotated blades. Besides determining whether a wind turbine is visible to the radio device, this method could also obtain which parts of the wind turbine are visible for partly visible wind turbines.

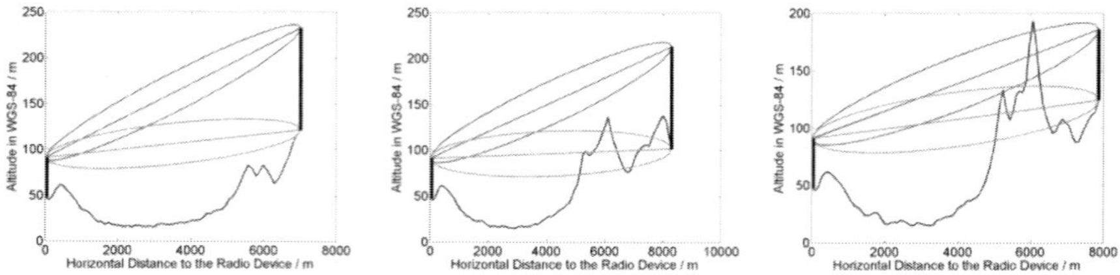

(a) Visible Completely (b) Visible Partly (c) Invisible Completely

Figure 2. Examples of Visibility Analysis Result.

Detailed Engineering Assessment

The working principles of different radio devices are quiet different, so we employ different detailed assessment method for different radio devices.

For PSR, the methods of simple engineering assessment and detailed engineering assessment in the impact assessment guidelines on surveillance sensors of Eurocontrol [2] could be utilized.

For SSR, which is the main surveillance device in civil aviation, we proposed an analysis method to calculate the range of wind farm's impact areas for detailed engineering assessment in the reference [16]. We amend some contents of the reference [16] in the following paragraphs. The SSR usually may be affected by the wind farm only when an aircraft flies in some certain airspace. We defined the projected areas of these airspaces as the impact areas. The impact areas are corresponding to four conditions as follows.

The first condition is that an aircraft and a wind turbine are both in the main lobe of the radar and the multipath response signal scattered by wind farms overlaps the direct response signal. In this condition the interference of the wind farm may cause decoding error, position error, decrease of the probability of detection, false target, track branching off or processor overload. This condition meets the formula as [16]

$$D_{rw} + D_{wa} - D_{ra} \le L_{th} \tag{7}$$

$$\frac{S}{I} = \frac{D_{rw}^2 D_{wa}^2}{D_{ra}^2} \frac{4\pi}{\sigma} < R_{th} \tag{8}$$

where D_{rw}, D_{ra}, D_{wa} are the slant distance form the radar to the wind turbine, from the radar to the aircraft and from the wind turbine to the aircraft respectively. S/I is the power radio of the direct response and the multipath response and σ is the bi-static RCS of the wind turbine. Path difference threshold L_{th} corresponds to the time delay threshold, which is 20.95μs (that is 20.3μs + 0.1μs + 0.45μs + 0.1μs because the time length of the response signal between F1 pulse and F2 pulse is 20.3μs ± 0.1μs and the width of F1 pulse is 0.45μs ± 0.1μs) in mode A and mode C. The value of 50dB could be used for R_{th} according to the reference [2].

The impact area for this condition is a sector in the main lobe of the radar as is shown in Figure 3. According to formula (7) the farthest horizontal distance to the radar of the impact area is infinite and nearest horizontal distance r_0 could be calculated by solving the equation

$$D_{rw} + D_{wa} - D_{ra} = L_{th} \tag{9}$$

If the equation above has no solution then the nearest horizontal distance is zero. Let d_{rw} be the horizontal distance from the radar to the wind turbine and h be the height of the aircraft corresponding to the radar as well as $k = R_{th}\sigma / 4\pi$. According to formula (8), we could obtain the nearest and the farthest horizontal distance to the radar of the impact area are

$$r_1 = \begin{cases} \dfrac{d_{rw} - \sqrt{k - \left(1-k/d_{rw}^2\right)^2 h^2}}{1-k/d_{rw}^2} & d > \sqrt{k/(1+\sqrt{k}/h)} \\ 0 & d \le \sqrt{k/(1+\sqrt{k}/h)} \end{cases} \tag{10}$$

$$r_2 = \begin{cases} \dfrac{d_{rw} + \sqrt{k - \left(1-k/d_{rw}^2\right)^2 h^2}}{1-k/d_{rw}^2} & d > \sqrt{k} \\ \infty & d \le \sqrt{k} \end{cases} \tag{11}$$

Then the final nearest horizontal distance is $\min(r_0, r_1)$ and the farthest horizontal distance is r_2 .

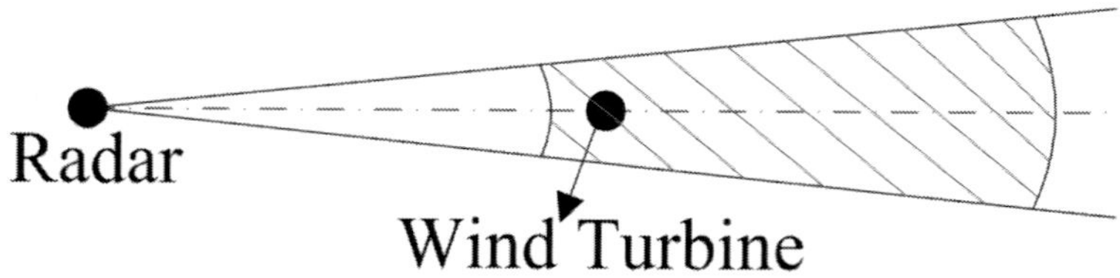

Figure 3. The Shape of the Response Impact Area (Shadowed Area).

The second condition is that an aircraft and a wind turbine are both in the main lobe of the radar and the P1 pulse of the multipath interrogation signal scattered by wind farms overlaps the P2 pulse of the direct interrogation signal. In this condition the direct interrogation signal may be regarded as a side lobe interrogation and be ignored. Then the probability of detection would decrease. This condition meets the formula

$$L_{th1} < D_{rw} + D_{wa} - D_{ra} < L_{th2} \tag{12}$$

The interval between P1 pulse and P2 pulse is 2μs±0.15μs and the width of P1 pulse and P2 pulse is 0.8μs±0.1μs. Therefore the path difference threshold L_{th1} and L_{th2} correspond to 0.95μs (2μs -0.15μs -0.8μs -0.1μs) and 3.05μs (2μs +0.15μs +0.8μs +0.1μs) respectively. The impact area for this condition has the same shape to that in Figure 3. The method to calculate the nearest and the farthest horizontal distance to the radar of the impact area is similar to the calculation procedure according to formula (7). The impact area for this condition is much smaller than that for the first condition.

The third condition is that a wind turbine is in the main lobe of the radar and the multipath interrogation signal has a large time delay and a large power. In this condition a false target may appear and the real aircraft is either in the main lobe or in the side lobe of the radar. Because ISLS (Interrogation Side Lobe Suppression) technique is adopted commonly, the false target may appear only when the time delay between the direct interrogation and the multipath interrogation is larger than 35μs. The third condition meets the formula [16]

$$D_{rw} + D_{wa} - D_{ra} > L'_{th} \tag{13}$$

$$P_{ref} = \frac{\sigma G_{rw} P_t G_{wa} \lambda^2}{4\pi\ ^{3}\ d_{rw}^2\ \ d_{wa}^2 + h^2} > P_{th} \tag{14}$$

where path difference threshold L'_{th} is 10.5 kilometers corresponding to 35μs and P_{th} is the sensitive of the receiver of the transponder which is usually -77dBm. In formula (14), P_{ref} is the power of the multipath interrogation signal; σ is the bi-static RCS of the wind turbine form the radar direction to the aircraft direction, which could be instead of the maximum bi-static RCS of the wind turbine; G_{rw} and G_{wa} are respectively the transmitter antenna gain of the radar and the receiver antenna gain of the transponder in the direction of the wind turbine; P_t is the transmitted power of the radar; λ is the wavelength of the electromagnetic wave; d_{rw} and d_{wa} are the horizontal distance between the radar and the wind turbine and between the wind turbine and the aircraft; h is the height of the aircraft corresponding to the radar.

The impact area for the third condition is shown in Figure 4. Corresponding to the formula (13) it is a part of a hyperbola near the wind turbine and corresponding to the formula (14) it is a part of a circle near the radar.

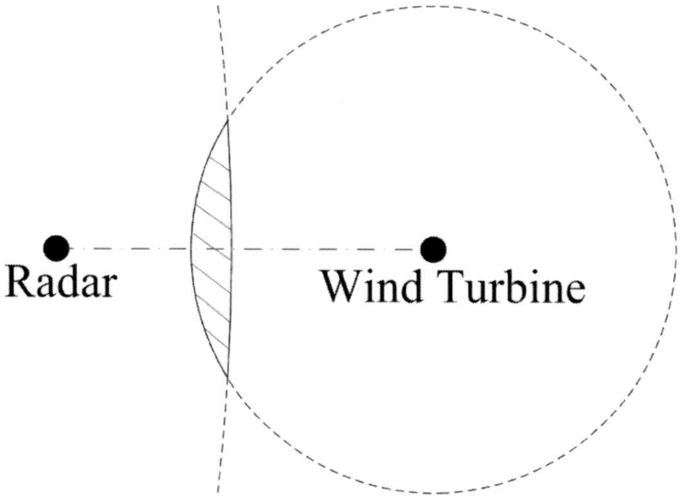

Figure 4. The Shape of the Impact Area for the Third Condition (Shadowed Area).

The fourth condition is an aircraft flies in the shadow of a wind turbine. The shadow airspace of a wind turbine could be calculated similar to PSR with the method in the annex A of impact assessment guidelines of Eurocontrol [2]. Then the impact area could be easily determined by the shadow airspace.

The impact areas of the wind farm are the summation of impact area corresponding to each wind turbine for above four conditions. The range of impact areas varies with several parameters such as the height of the aircraft and the RCS of the wind turbine. It is discussed in the reference [16].

For VOR, we only analyze the impact on DVOR for DVOR is commonly utilized currently. We calculate the azimuth angle errors of different aircraft positions with [17]

$$e = \arctan\left(\frac{\sum_{i=1}^{N} 2\rho_i J_1\left[2n\sin \phi'_i/2\right]\cos \phi'_i/2}{n-\sum_{i=1}^{N} 2\rho_i J_1\left[2n\sin \phi'_i/2\right]\cos \phi'_i/2}\right) \tag{15}$$

and then get the maximum azimuth angle error. In formula (15), N is the number of the wind turbines; n is the frequency modulation index of the signal; J_1 is the one order Bessel function. Let ϕ_0 be the azimuth angle of the aircraft and ϕ_i be the azimuth angle of the ith wind turbine corresponding to the VOR, then ϕ'_i in formular (15) equals $\phi_i - \phi_0$. ρ_i in formular (15) could be calculated by $\rho_i = a_i \cos \theta_i - \theta_0$ where a_i is amplitude radio of the direct signal and the multipath signal scattered by the ith wind turbine; θ_0 is the phase of the direct signal; θ_i is the phase of the multipath signal scattered by the ith wind turbine. In the calculation we only consider the extreme cases that $\cos \theta_i - \theta_0$ equal 1 or -1. a_i is determined by the path difference between the direct signal and the multipath signal and the bi-static RCS of the ith wind turbine. The RCS in a particular direction or the maximum RCS could be utilized for the calculation of a_i.

For the localizer and the glide slope of ILS, the impact of the wind farm on the difference of degree of modulation (DDM) and corresponding position error are calculated to assess the impact of the wind farm.

3D Visualized Assessment Based on Geographic Information System

In order to represent the assessment results in an explicit style, we proposed a 3D visualized assessment method for impacts of wind farms on surveillance and navigation radio devices in civil aviation based on geographic information system [18]. The method could be employed to represent the assessment results of zone partition assessment and visibility assessment mentioned above.

The method calculates the longitude, latitude and altitude of points on the boundaries of different types of partitioned zones for zone partition assessment according to related parameters of the radio devices and wind turbines. The longitude, latitude and altitude of points on the boundaries of the scope of line of sight from the radio device to wind turbines are also calculated for visibility assessment. The positions of key points are usually given with distance, height and azimuth angle corresponding to the radio device. Therefore, the method of forward solution of geodetic problem with the spherical earth model is utilized to obtain the longitude, latitude and altitude of key points. The formulas for forward solution are

$$\sigma = d / R \tag{16}$$

$$\sin\phi_2 = \sin\phi_1 \cos\sigma + \cos\phi_1 \sin\sigma \cos\alpha_1 \tag{17}$$

$$\tan\lambda = \frac{\sin\sigma\cos\alpha_1}{\cos\phi_1\cos\sigma - \sin\phi_1\sin\sigma\cos\alpha_1} \tag{18}$$

$$\lambda_2 = \lambda_1 + \lambda \tag{19}$$

where α_1 is the azimuth angle of point 1 corresponding to point 2, which has some additional difference except for 180 degree difference compared with the azimuth angle of point 2 corresponding to point 1. Here α_1 is the azimuth angle of the wind turbine corresponding to the analyzed radio device, which is the same to α in formula (3). The meaning of other symbols in formula (16) to (19) is the same to that in formula (3).

Then we record the different types of partitioned zones and the scope of line of sight in a KML (Keyhole Markup Language) file with points on boundaries. KML file is an XML file for expressing geographic annotation and visualization of geographic information. We write the coordinates of points on boundaries into a file with a program automaticly following the specification of the KML file. The transparency and color of added graphics also could be set in a KML file. The transparency is important for the method becauce it make the partitioned zones and visualized geographic information could be seen at the same time.

(a) Zone Partition Assessment for a Localizer of ILS

(b) Visibility Assessment (the View near the Radar)

Figure 5. Examples of 3D Visualized Assessment Results.

Figure 6. Distribution of a Planned Wind Farm Observed from the Direction of the Planned Radar.

Finally the KML file which represnets the assessment results could be loaded in a 3D geographic information such as Google Earth. Then we could obtain the 3D visualized assessment results. These results combined with the geographic information such as terrain and ground objects are more explict. Some examples of 3D visualized assessment results are shown in Figure 5. Meanwhile, the geographic position relationship of the radio device and the wind farm could be displayed explicitly in a 3D style including terrain information with this method. An example is shown in Figure 6. This is very useful for decision-making of planned project.

Impact Assessment Software

We developed a sort of software for visualized assessment based on the 3D visualized assessment method mentioned above. We can obtain the 3D visualized assessment result for zone partition assessment and visibility assessment through loading the parameters of radio devices and wind turbines with this software. A user interface of this software is shown in Figure 7.

Figure 7. A User Interface of the 3D Visualized Assessment Software.

Application of the Impact Assessment

There are some cases that the planned radio devices in civil aviation near the existed or the planed wind farm in China. Some of our assessment methods have been utilized in the actual cases of impact assessment of the wind farm in Shandong and Chongqing of China.

RCS Estimation of a Wind Turbine

The proper estimation of the RCS of a wind turbine is very important to quantitative detailed engineering assessment. The RCS varies obviously with several factors such as the direction of the incident electromagnetic wave, the direction of the scattering, the frequency of the incident electromagnetic wave, the geometrical size of the wind turbine, the material of the wind turbine, the shape of blades, the direction of the blades' rotation plane, the blades' position, the blades' tilt angle, etc. Although some RCS measurement experiments for real wind turbines have been made [9][19], it is very difficult to measure the all possible cases and it is impossible for planned wind farms. Researchers from University of Oklahoma have made a scaled model of the wind turbine to measure the RCS of different angle and different polarization in the laboratory and to analysis the scattering characterization of the wind turbine [20-22]. That is a good idea to estimate the RCS of a wind turbine. Other researchers utilized a CAD model of the wind turbine and calculated the RCS of a wind turbine with electromagnetic calculation methods [23].

We employ electromagnetic calculation methods for RCS estimation in our impact assessment. There are several different electromagnetic calculation methods such as method of moments (MOM), fast multipole method (FMM), finite difference time domain (FDTD), finite element method (FEM), physical optics (PO), geometrical optics (GO), physical theory of diffraction (PTD), geometrical theory of diffraction (GTD). Considering the trade-off between accuracy and efficiency of the calculation, we mainly employ physical optics to estimate the RCS of a wind turbine.

In order to calculate the RCS of a wind turbine, we separate the wind turbine into three parts: tower, nacelle and blades. Because the geometrical size and the RCS of the nacelle are both smaller than other two parts commonly, we ignore the scattering of the nacelle.

The tower of the wind turbine is usually made of steel and could be modeled as a perfect electric conductor (PEC). The shape of the tower is a truncated cone. Although the difference of radius between the top and the bottom of the tower is small, we find the difference of RCS between a truncated cone and a cylinder with similar size is obvious for different directions. Therefore the tower should be modeled as a truncated cone in the electromagnetic calculation.

The blades are commonly made of complex material of glass fiber and epoxy resin and could be modeled as a dielectric whose dielectric constant is between 4 and 5. The shape of the blade is very complex. The precise blade model could be depicted with the blade element model. The blade element model depicts the parameters such as the distance from the root, chord, twist, twist axis, thickness, pitch axis, pre-bend, aero-dynamic-control and aerofoil section reference of each section element along the blade. It is very complex to model a CAD model with the blade element model. In addition, the data of the blade element model is difficult to be obtained in the actual impact assessment cases.

In order to simplify the RCS estimation of blades, we proposed a parameterized blade model in the reference [24]. We extract some key parameters of a blade for our parameterized blade model. These key parameters include blade length, number of blade elements, maximal chord length, diameter of root, position of the blade element with maximal chord length, twist angle of the root element and maximal twist angle. Then the CAD model could be modeled with these parameters. The RCS of blades in different azimuth angle and different pitch angle calculated by the precise blade element model and by the parameterized blade model are compared. The result of the compare indicates that the RCS of blades calculated by the parameterized blade model is similar to that calculated by the precise blade element model. The parameterized blade model is sufficient for the impact assessment. The variation of the RCS of blades with different key parameters is further analyzed. The further analysis shows that the RCS of a blade is not sensitive to the key parameter maximal chord length, but it is sensitive to the key parameter diameter of root and maximal twist angle. Larger diameter of root corresponds to larger maximal RCS. Smaller maximal twist angle corresponds to larger maximal RCS in the vertical plane but almost has no influence in the horizontal plane.

A blade could be further simplified with a cylinder. We also compare the RCS of blades calculated by the precise blade element model and calculated by the simplified cylinder blade model in the reference [25]. The analysis indicates that the RCS results of blades in different azimuth angel and pitch angle calculated by two models have obvious difference. The RCS estimation with the simplified cylinder blade model would get larger errors. However, the largest RCS results of blades calculated by two models are similar. Therefore, the simplified cylinder blade model could be utilized to coarsely estimate the maximal RCS of blades.

CONCLUSION

The potential impacts of wind farms on radio devices in civil aviation are summarized in this chapter. In addition, a review of the assessment procedure and methods for the impacts of wind farms and the RCS estimation methods for a wind turbine of our research group are presented.

The contents of this chapter could be applied to several aspects. 1) For establishing a new surveillance or navigation radio device on the ground near a wind farm, the impact assessment could provide technical supports for the site selection and the equipment selection of the radio device. 2) For planning a new wind farm near radio devices of civil aviation, the impact assessment could provide technical supports for scheme establishing of the wind farm. 3) For established surveillance or navigation radio device near an existed wind farm, the impact assessment could analysis the potential impact. It is very valuable for assessing the safety risk for civil aviation as well as the selection and implementation of interference mitigation measures. Therefore it is of great importance for the safeguarding of the flight of civil aviation.

ACKNOWLEDGMENTS

The authors would like to thank the supports of the National Science Foundation of China (No. U1233109, No. U1533110) and the National University's Basic Research Foundation of China (No. 3122015D005).

REFERENCES

[1] UK Civil Aviation Authority. (2013). *CAA policy and guidelines on wind turbines* (5th ed.). London: The Stationery Office.

[2] Eurocontrol. (2014). *Eurocontrol guidelines on how to assess the potential impact of wind turbines on surveillance sensors* (1.2 ed.). Brussels: Eurocontrol Headquarters.

[3] Lemmon, J. J., Carroll, J. E., Sanders, F. H. & Turner, D. (2008). *Assessment of effects of wind turbines on ATC radars.* Boulder, CO: NTIA/ITS Technical Publications Office.

[4] Radio Advisory Board of Canada, & Canadian Wind Energy Association. (2007). *Technical information and guidelines on the assessment of the potential impact of wind turbines on radiocommunication, radar and seismoacoustic systems.* Retrieved from http://citeseerx.ist.psu.edu/viewdoc/download?doi=10.1.1.560.447&rep=rep1&type=pdf.

[5] Theil, A., Schouten, M. W. & de Jong, A. (2010). Radar and wind turbines: A guide to acceptance criteria. In *Radar Conference*, 2010 *IEEE*, (pp. 1355-1361). IEEE.

[6] Angulo, I., de la Vega, D., Cascón, I., Cañizo, J., Wu, Y., Guerra, D. & Angueira, P. (2014). Impact analysis of wind farms on telecommunication services. *Renewable and Sustainable Energy Reviews*, *32*, 84-99.

[7] De la Vega, D., Fernandez, C., Grande, O., Angulo, I., Guerra, D., Wu, Y. & Ordiales, J. L. (2011). Software tool for the analysis of potential impact of wind farms on radiocommunication services. In *Broadband*

Multimedia Systems and Broadcasting (BMSB), 2011 IEEE International Symposium on, (pp. 1-5). IEEE.

[8] Office of the Director of Defense Research and Engineering. (2006). *The effect of windmill farms on military readiness*. Retrieved from http://citeseer.ist.psu.edu/viewdoc/download?doi=10.1.1.127.1771&rep=rep1&type=pdf.

[9] Rudd, R. & Rhandawa, B. (2009). RCS measurement of wind turbines. In *Antennas and Propagation, 2009 3rd European Conference on*, (pp. 3642-3644). IEEE.

[10] Wang, X., Ma, Y., He, W., Wang, W. & Wu, R. (2015). An assessment of wind farms' electromagnetic impact for the aerodrome. In *Integrated Communication, Navigation, and Surveillance Conference (ICNS), 2015* (pp. H4-1 - H4-10). IEEE.

[11] Lute, C. & Wieserman, W. (2011). ASR-11 radar performance assessment over a wind turbine farm. In *Radar Conference*, 2011 *IEEE*, (pp. 226-230). IEEE.

[12] Odunaiya, S. A. (2006). Wind farms and their effect on radio navigation aids. In *Proceedings of 14th SIIV IFIS*, (pp. 77-80).

[13] Greving, G. (2004). Modern threats to precision approach and landing – the A380 and windgenerators and their adequate numerical analysis. In *Precision Approach and Landing (ISPA)*, 2004 *International Symposium on*, (pp. 1-12).

[14] International Civil Aviation Organization (ICAO). (2015). *European guidance material on managing building restricted areas* (3rd ed.). Retrieved from http://www.icao.int/EURNAT/EUR and NAT Documents/EUR Documents/015 - Building Restricted Areas/ICAO EUR Doc 015 Third Edition Nov2015.pdf.

[15] Wang, X. (2013). Visibility analysis of wind turbines applied to assessment of wind farms' electromagnetic impact. In *Information Science and Technology (ICIST), 2013 International Conference on*, (pp. 1277-1280). IEEE.

[16] Wang, X., Ma, Y., He, W., Xu, M. & Wu, R. (2015). Analysis on the range of wind farm's impact area for SSR. In *2015 IET International Radar Conference*. IET.

[17] Morlaas, C., Fares, M. & Souny, B. (2008). Wind turbine effects on VOR system performance. *Aerospace and Electronic Systems, IEEE Transactions on*, *44*(4), 1464-1476.

[18] Wang, X., Yue, S. & Wu, R. (2015). 3D visualized assessment for wind farms impact on surveillance and navigation in civil aviation. *Journal of Signal Processing*, *31*(10), 1307-1312. (in Chinese).

[19] Poupart, G. J. (2003). *Wind farms impact on radar aviation interests–final report*. Retrieved from http://webarchive.nationalarchives.-gov.uk/+/http://www.berr.gov.uk/files/file42919.pdf.

[20] Zhang, Y., Huston, A., Palmer, R. D., Albertson, R., Kong, F. & Wang, S. (2011). Using scaled models for wind turbine EM scattering characterization: techniques and experiments. *Instrumentation and Measurement, IEEE Transactions on*, *60*(4), 1298-1306.

[21] Kong, F., Zhang, Y., Palmer, R. & Bai, Y. (2011). Wind turbine radar signature characterization by laboratory measurements. In *Radar Conference, 2011 IEEE*, (pp. 162-166). IEEE.

[22] Kong, F., Zhang, Y. & Palmer, R. D. (2013). Wind turbine radar interference studies by polarimetric measurements of a scaled model. *Aerospace and Electronic Systems, IEEE Transactions on*, *49*(3), 1589-1600.

[23] Kent, B. M., Hill, K. C., Ugh, A. B., Zelinski, G., Hawley, R., Cravens, L. & Coveyou, T. (2008). Dynamic radar cross section and radar Doppler measurements of commercial General Electric windmill power turbines Part 1: Predicted and measured radar signatures. *Antennas and Propagation Magazine, IEEE*, *50*(2), 211-219.

[24] Ma, Y., Sun, J., Wang, X., He, W., Wang, W. & Wu, R. (2015). Simulating RCS of the parameterized blades of a wind turbine. In *Integrated Communication, Navigation, and Surveillance Conference (ICNS), 2015* (pp. O4-1 - O4-9). IEEE.

[25] Ma, Y., Wu, R., Wang, X. & He, W. (2015). Full analysis of RCS of wind turbine blades. In *2015 IET International Radar Conference*. IET.

BIOGRAPHICAL SKETCH

Name: Xiaoliang WANG

Affiliation: Tianjin Key Laboratory for Advanced Signal Processing, Civil Aviation University of China, Tianjin, P.R.China

Date of Birth: August 16th, 1982

Education:

Xiaoliang WANG received the B.S. degree in electronics and information engineering from Communication University of China, Beijing, China, in 2004 and the Ph.D. degree in information and communication engineering from Beihang University, Beijing, China, in 2010.

Address: Civil Aviation University of China, No. 2898, Jinbei Road, Dongli District, Tianjin, 300300, China.

Research and Professional Experience:

From 2004 to 2010 Xiaoliang WANG was a postgraduate in School of Electronics and Information Engineering, Beihang University, Beijing, China and did research on image processing and recognition. He is currently an instructor in Civil Aviation University of China (CAUC), Tianjin, China. Since coming to CAUC after receiving his Ph.D. degree in 2010, He began to do research on communication, navigation and surveillance techniques in civil aviation. His research interests have been focused on the assessment and mitigation of impacts caused by wind farms on civil aviation and the supporting technique for the operation of general aviation in recent five years.

Professional Appointments:

He has been an instructor in Civil Aviation University of China (CAUC), Tianjin, China from 2010.

Honors:

- The Best Safe and Secure Air Transportation System Paper in Integrated Communication, Navigation, and Surveillance Conference (ICNS), 2015.
- The 3rd Place Professional Paper in Integrated Communication, Navigation, and Surveillance Conference (ICNS), 2015.

Publications Last 3 Years:

1. Wang, X., Ma, Y., He, W., Wang, W., & Wu, R. (2015). An assessment of wind farms' electromagnetic impact for the aerodrome. In *Integrated Communication, Navigation, and Surveillance Conference (ICNS), 2015* (pp. H4-1 - H4-10). IEEE.
2. Wang, X., Ma, Y., He, W., Xu, M., & Wu, R. (2015). Analysis on the range of wind farm's impact area for SSR. In *2015 IET International Radar Conference*. IET.
3. Wang, X., Yue, S., & Wu, R. (2015). 3D visualized assessment for wind farms impact on surveillance and navigation in civil aviation. *Journal of Signal Processing, 31(10)*, 1307-1312. (in Chinese)
4. Wang, X. (2013). Visibility analysis of wind turbines applied to assessment of wind farms' electromagnetic impact. In *Information*

Science and Technology (ICIST), 2013 International Conference on (pp. 1277-1280). IEEE.

5. Wang, X., Ma, Y., Wang, P., & Wu, R. (2015). Design and implementation of flight plan acceptance system for general aviation. *Computer Engineering and Design, 36(10)*, 2838-2842 & 2848. (in Chinese)
6. Ma, Y., Sun, J., Wang, X., He, W., Wang, W., & Wu, R. (2015). Simulating RCS of the parameterized blades of a wind turbine. In *Integrated Communication, Navigation, and Surveillance Conference (ICNS), 2015* (pp. O4-1 - O4-9). IEEE.
7. Ma, Y., Wu, R., Wang, X., & He, W. (2015). Full analysis of RCS of wind turbine blades. In *2015 IET International Radar Conference*. IET.
8. He, W., Ma, Y., Shi, Y., Wang, X., Zhang, S., & Wu, R. (2015). Analysis of the wind turbine RCS and micro-Doppler feature based on FEKO. In *Integrated Communication, Navigation, and Surveillance Conference (ICNS), 2015* (pp. U3-1 - U3-9). IEEE.
9. He, W., Zhai, Q., Wang, X., Zhang, S., & Wu, R. (2015). Wind turbine clutter detection in scanning ATC radar. In *Integrated Communication, Navigation, and Surveillance Conference (ICNS), 2015* (pp. U4-1 - U4-8). IEEE.
10. He, W., Shi, Y., Ma, Y., Wang, X., & Wu, R. (2015). Wind turbine clutter mitigation based on matching pursuits. In *2015 IET International Radar Conference*. IET.
11. He, W., Zhai, Q., Wang, X., Ma, Y., & Wu, R. (2015). Mitigation of wind turbine clutter based on the periodicity in scanning ATC radar. In *2015 IET International Radar Conference*. IET.
12. He, W., Guo, S., Wang, X., & Wu, R. (2015). Micro-Doppler features analysis of wind farm echoes for air traffic control radar in scanning mode. *Journal of Signal Processing, 31(10)*, 1240-1246. (in Chinese)
13. He, W., Shi, Y., Wang, X., Ma, Y., & Wu, R. (2015). Simulation and analysis of wind turbine echoes. *Journal of System Simulation, 27(1)*, 50-56. (in Chinese)
14. Wu, R., Mao, J., Wang, X., & Jia, Q. (2013). Target detection of primary surveillance radar in wind farm clutter. *Journal of Electronics and Information Technology, 35(3)*, 754-758. (in Chinese)
15. Wu, R., Fu, H., Wang, X., & Jia, Q. (2013). Impact assessment of wind farms on secondary surveillance radar utilizing signal

characteristics. *Journal of Civil Aviation University of China*, *31(6)*, 1-4. (in Chinese)
16. Chen, M., Wang, X., Wang, W., & Wu, R. (2015). High precision PSR data simulation method based on non-equal interval sampling. *Computer Simulation*, *32(2)*, 111-114. (in Chinese)
17. Wang, P., Wang, X., Zhang, Z., & Wu, R. (2015). General aviation fight data processing system based on stored procedure. *Computer Engineering and Design*, *36(4)*, 1084-1089. (in Chinese)
18. Wu, R., Wang, P., & Wang, X. (2014). Process design of general aviation synthetic operation support. *Journal of Civil Aviation University of China*, *32(3)*, 1-5. (in Chinese)
19. Wu, R., Liu, Y., & Wang, X. (2014). Implementation of track estimation based on flight plan for general aviation. *Journal of Civil Aviation University of China*, *32(1)*, 1-4 & 9. (in Chinese)

INDEX

D

E

F

G

H

I

X

Y